LIFE CYCLE ASSESSMENT

LIFE CYCLE ASSESSMENT

Principles, Practice and Prospects

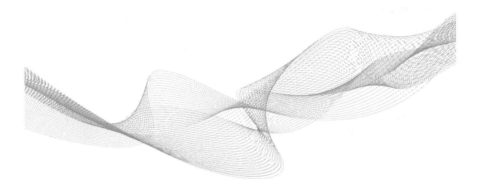

Ralph Horne • Tim Grant • Karli Verghese

CSIRO
PUBLISHING

National Library of Australia Cataloguing-in-Publication entry
Horne, Ralph 1966–

Life cycle assessment : principles, practice and prospects
/Ralph Horne, Tim Grant, Karli Verghese.

9780643094529 (pbk.)

Includes index.
Bibliography.

Environmental impact analysis.
Environmental risk assessment.

Grant, Tim.
Verghese, Karli.

333.714

Published by
CSIRO PUBLISHING
150 Oxford Street (PO Box 1139)
Collingwood VIC 3066
Australia

Telephone: +61 3 9662 7666
Local call: 1300 788 000 (Australia only)
Fax: +61 3 9662 7555
Email: publishing.sales@csiro.au
Web site: www.publish.csiro.au

Front cover image by iStockphoto

Set in Adobe Minion 10/12 and Stone Sans
Cover and text design by James Kelly
Typeset by Desktop Concepts Pty Ltd, Melbourne
Printed in Australia by Ligare

The book has been printed on paper certified by the Programme for the Endorsement of Forest Chain of Custody (PEFC). PEFC is committed to sustainable forest management through third party forest certification of responsibly managed forests.

CSIRO PUBLISHING publishes and distributes scientific, technical and health science books, magazines and journals from Australia to a worldwide audience and conducts these activities autonomously from the research activities of the Commonwealth Scientific and Industrial Research Organisation (CSIRO).

The views expressed in this publication are those of the author(s) and do not necessarily represent those of, and should not be attributed to, the publisher or CSIRO.

Contents

Foreword

Our choices now are deciding the broad direction of our future. Our aim must be lifestyles that would be sustainable. A series of scientific studies at local, national and global level all tell the same story: we are over-using the natural resources of the planet and stressing its natural systems. Climate change is the most urgent priority, but it is only one of several indications that our present lifestyle is unsustainable. We must stop consuming the resources our children will need. We must also stop destroying the natural wonders of this world. A sustainable future will involve fundamental changes.

We have to make responsible choices about the essentials of a civilised life: food, shelter, water and waste management. Intuition is not a reliable guide, neither is superficial analysis. We need sophisticated Life Cycle Assessment. LCA is systematic analysis considering all steps of the process of using natural resources to provide our needs. It also assesses the impacts of by-products of that process. Whether you are making decisions as an individual, a householder or a manager of a commercial activity, you need the sort of informed basis this book provides. As well as explaining the techniques of Life Cycle Assessment, it also discusses its limitations. So it is a timely contribution to the most urgent task for our generation, modifying our lifestyle to achieve human needs within the limits of natural systems.

Ian Lowe
Emeritus Professor of Science, Technology and Society, Griffith University
Adjunct Professor, Sunshine Coast University and
Queensland University of Technology
Honorary Research Fellow, University of Adelaide
Officer of the Order of Australia

Acknowledgements

Life Cycle Assessment: Principles, Practice and Prospects includes the synthesis of many years' work by the authors, all of which has been undertaken in interdisciplinary team environments. We would like to acknowledge the indirect input of our many collaborators, mentors, supporters and associates, who are too numerous to name here in full. A shorter list would include Nigel Mortimer, Robert Matthews, Frances Pamminger, Tony Kelly, Helen Lewis, Kees Sonneveld, Greg Norris, Mark Goedkoep, Bo Weidema, Olivier Jolliet and Frank Fisher for their various significant influences, inspirational work in the field of LCA and perspectives on systems thinking.

The authors also wish to acknowledge all those who made direct contributions to the book, including Fran McDonald who provided essential subediting and comments on the text throughout. We are also indebted to those who contributed to specific chapters, including Rob Rouwette, who provided input and advice on the Dutch case study in Chapter 7; Dominique Hes, who provided input and advice on the Sydney Olympics case study; and Andrew Walker-Morison, who conducted development work on the BAMS project also mentioned in Chapter 7. Annette Cowie provided input on issues related to forest carbon and LCA in Chapter 10, while in Chapter 11 John Gertsatkis and Kendra Wasiluk contributed to ideas and projects reported here by the authors. In acknowledging these contributions, we would also emphasise that any errors or omissions are entirely ours.

Ralph Horne
Tim Grant
Karli Verghese
November 2008

Introduction

Life cycle assessment (LCA) has developed rapidly over recent decades into a technique for systematically identifying the resource flows and environmental impacts associated with the provision of products and services. Interest in LCA has accelerated alongside growing demand to assess and reduce greenhouse gas emissions across different manufacturing and service sectors.

Life Cycle Assessment: Principles, Practice and Prospects is designed to provide practitioners or interested professionals with a critical insight into the technique of LCA. Instead of being a handbook or a solely theoretical text, the book focuses on the reflective practice of LCA, in other words, how LCA practice has developed and must continue to develop empirically, through practice and reflection on this practice. In so doing it also provides academics, scholars and students with a journey into reflective practice.

The early chapters describe the distinctive strengths and limitations of LCA, with an emphasis on practice in Australia. The second half of the book includes chapters on the application of LCA to the key sectors of waste management, the built environment, water and agriculture. In each chapter, the contemporary challenges for environmental assessment and performance improvement in these sectors are investigated, well illustrated with examples and case studies. The last three chapters project current issues and debates to 2020 and beyond. LCA methodologies are compared to the emerging climate change mitigation policy and practice techniques, and the uptake of 'quick' LCA and management tools are considered in the light of current and changing environmental agendas. In the last chapter, the authors debate the future prospects for LCA technique and applications.

Chapter 1

Life cycle assessment: origins, principles and context

Ralph E Horne

The launch of the International Organization for Standardization's ISO 14001 ('Environmental management systems – Specification with guidance for use') in 1996 indicated to many businesses that *ad hoc* environmental management was no longer an option. For an increasing number of organisations, regulations, business drivers and the public environmental and social concerns had reached a level where a more strategic and systematic approach to environmental challenges was necessary. The resultant rapid rise in corporate environmental management and accompanying discourse is already well charted and critiqued (e.g. Rondinelli and Vastag 2000; Sandström 2005).

Contemporaneously, life cycle assessment (LCA) began to produce convincing evidence that intuition was no longer enough either. 'Natural' products were found to be not necessarily environmentally optimal. Many 'counter-intuitive' outcomes from LCA studies indicated the need for a closer systemic approach to identify and document impacts along the process chain and life time of products and services. Business began to take a greater interest in LCA (Frankl and Rubik 2000) and a series of texts appeared on the subject (e.g. Curran 1996; Ciambrone 1997; Graedel 1998). The physical sciences and engineering disciplines began to recognise LCA as a tool to help reconcile values, technological impacts and the environment (Hofstetter 1999) and the United Nations (UN) began to envisage the global roll-out of LCA practice (UNEP 2000).

As the roll-out gathers pace, this book is intended to provide scholars and professionals across a range of disciplines with a critical perspective on the practice of LCA and its possible future directions. It is not intended as a guide or handbook, of which there are several already (e.g. Baumann and Tillman 2004). Instead, theory, methodologies and applications of LCA are critically examined. Key developments, challenges and opportunities are illustrated with case studies.

This chapter, as well as introducing subsequent chapters, charts the origins of LCA and then defines the technique before placing it within the wider context of environmental management and assessment.

1.1 LCA origins

Humans have long been concerned with the energy efficiency of technologies and the services they provide. Perennial questions arise from Newton's First Law of Thermodynamics – if energy is never lost, in what proportions does it dissipate through various processes? What is the energy benefit and loss in various processes? Also, specifically, for energy 'generation' (i.e. 'conversion'), how much input energy is necessary to produce a given energy service? Then again, and perhaps most topically, what are the implications for climate change of different energy scenarios, and how can we identify optimal services from 'sustainable' levels of impact?

In the post-World War 2 era, a new generation of energy technologies – nuclear, geothermal, modern wind and other 'renewables' – tested the energy balance question. Energy analysis became increasingly complex, systemic and sophisticated through successive empirical developments. At first, the technique was typically used to assess production of a given unit of energy in a given form, examining immediate inputs on site in the production system. Inevitably, as more complex generation technologies were examined, so the analysis was extended. For example, the question of whether a given nuclear generation technology produced more energy than it consumed led researchers to look beyond the generation facilities themselves to 'yellow cake' production and uranium mining, long-term waste management, and even to the impacts of transport (of personnel, materials and equipment) and associated research, development, marketing and management services. This was a precursor to what became known as 'life cycle analysis', a systematic process-oriented approach to identifying energy inputs in the production of energy services.

In the late 1960s, the first Resource Environmental Profile Analyses (REPAs) were undertaken and these became the forerunners of modern LCA. Notably, Coca Cola Amatil commissioned work by a group of researchers (later to become Franklin Associates) in the United States of America (USA) to investigate the resources and environmental profile of different packaging materials for their products (Hunt *et al.* 1974). Oil shortages in the early 1970s led to a focus back on energy analysis. However, by the mid-1980s, multi-criteria systematic inquiry had spread to include nappies, appliances, automobiles and housing. Interchangeable terms were used to describe these studies including eco-balances, cradle-to-grave analysis, and life cycle analysis. In 1990, the term 'life cycle assessment' was proposed and agreed upon by those attending a workshop in Vermont, USA, held by the Society of Environmental Toxicology and Chemistry (SETAC).

Rapid development followed, as LCA grew into a body of systematic, inclusive, analytical approaches to environmental impact assessment. SETAC then embarked on the development and extension of LCA, publishing various 'best-practice' guides (e.g. Barnthouse *et al.* 1997; Kotaji *et al.* 2002; Udo de Haes *et al.* 2002) and advice on LCA simplification and methods (e.g. Udo de Haes 1996; Christiansen 1997). Applications to public policy (Allen *et al.* 1997) and particular sectors, such as building and construction (Kotaji *et al.* 2003), were also examined as well as the application of LCA to more embedded management modes within organisations (Hunkeler *et al.* 2004).

In 2002, the United Nations Environment Programme (UNEP) and SETAC formed the UNEP/SETAC Life Cycle Initiative to assist development and uptake of LCA. Building upon a base of practice in several European countries, the USA and Japan, this initiative seeks to enable users to put life cycle thinking into practice. This has generated focus on the 'new' manufacturing centres in Asia (including the subcontinent and eastern Asia), Africa and South America. Hence, as the production centres of modern manufacturing shifted through the effects of globalisation, so LCA practitioners followed, further developing methods and techniques for calculating the environmental impacts of production and consumption systems. The apparent interdependence of evolving LCA practice and demand across geographic regions has specific implications in Australia.

1.2   LCA definition, standards and process

Although many definitions exist, LCA essentially comprises a systematic evaluation of environmental impacts arising from the provision of a product or service. The original Interna-

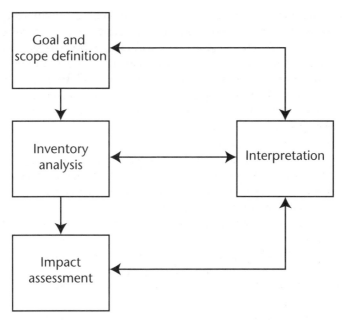

Figure 1.1   Outline of generic life cycle assessment (LCA) process (after ISO 14040, 2000a).

tional Organization for Standardization (ISO) definition provides some indication, although it is self-referencing: 'compilation and evaluation of the inputs and outputs and the potential environmental impacts of a product system throughout its life cycle' (ISO 1997, p. 2). Generic LCA method requires that all the main inputs to the processes that provide the service are taken into account, as well as the processes and materials that feed into those processes, and so on back 'up' the supply chains of the various materials in the product to the raw resource inputs. These raw inputs are invariably energy-based – the coal mine or oil well – rather than simply raw materials. For example, making bricks may require brick clay and an extraction quarry, but this process operates with fossil fuel-powered machinery. Hence, although bricks are made from quarried clay and other materials, at the end of this process is the oil well or coal mine required to drive the steel mill to make the machinery used in clay extraction.

International standards assist in the specification, definition, method and protocols associated with undertaking, reviewing and reporting LCA studies. ISO 14040 describes the principles and framework for life cycle assessment. The original standard (produced in 1997) was updated in 2006 (ISO 2006a). This 'core' standard includes guidance on defining the goal and scope of an LCA study, development of the life cycle inventory, the life cycle impact assessment, and interpretation (Fig. 1.1). It also indicates reporting and critical review parameters and limitations of LCA. However, it does not describe the LCA technique in detail, nor does it specify how to undertake individual phases of the LCA. More detail is provided in ISO 14044 (ISO 2006b), which together with ISO 14040:2006 replaces other former LCA-related standards (ISO 14040:1997, ISO 14041:1999, ISO 14042:2000 and ISO 14043:2000). Further standards deal with issues such as data documentation formats. Additional guides to the standards seek to provide more detail on their application in practice (e.g. Guinee 2002). However, not all regions have adopted the updated standards at the time of writing (e.g. Australia and New Zealand, AS/NZS ISO 1998).

For any LCA, appropriate framing of the key 'question' forms part of the definition of the goal and scope, including setting the functional units of the study. For example, a comparative

LCA of two coffee machines may indicate that the more durable, heavily built of the two has the higher environmental impact, if the product comparison is based on the product level. If, however, it is based on the functional level, we may find that the more durable product has a life-span which enables it to produce five times as many cups of coffee over its lifetime. Notwithstanding any functional differences in coffee quality, aesthetic quality, obsolescence or maintenance, this quality alone may reverse the outcome of the LCA comparison, simply by taking as the functional unit 'impact per production of 10 000 cups of coffee' rather than 'impact per coffee machine'.

The most well-known application of LCA is in comparing the 'total' environmental impact of a product or service with an alternative (comparable) product or service. UNEP refers to LCA as a tool to reveal 'the world behind the product' (Fava 2002). Hence, LCA is often considered a tool that provides 'the answer' to the question of which product has least environmental impact. However, LCA can reveal things other than the answer. It can also fail to reveal the answer at all, if the question is not precisely and appropriately framed (see Chapter 4 for further discussion of this point).

Defining the scope involves determining the appropriate limits of the analysis. This includes identifying the entire production and disposal or recycling process of the materials and services involved in the life cycle of the product or service being studied (and any comparative product or service). The components involved in delivering the product or service should be included, as well as all inputs to those components, and the inputs to those inputs, and so on. It also includes the outputs, emissions and wastes produced at all stages of the product or service delivery – both 'pre-consumption' and 'post-consumption'. Decisions may be taken to 'truncate' the system for practical purposes, and quick estimates of impacts more distant from the central processes may be undertaken to check that they are negligible and can be disregarded from a detailed assessment.

The resultant 'process chains' in the products or services under comparison may be significantly different. For example, a wool carpet and a synthetic carpet (for which an appropriate functional comparison might be 'the provision of 1 m^2 of carpet for 10 years') would have very different process chains, one being dominated by agricultural inputs and processes, the other by industrial ones. This example also raises the issue of allocation of impacts; while sheep farming produces wool, it also produces other animal products and the total impact of sheep farming is therefore only partly attributable to wool, with the remainder attributable to meat, hide and other sheep farming products.

The inventory is the result of compiling all environmental 'flows', including resource use inputs and waste or pollution outputs. This inventory provides a lower estimate of the environmental burdens that the product or service places upon the environment. However, the relative importance of these burdens requires some measure or indicator of impact. Inventory data can only be converted into impact results through the use of appropriate algorithms or indicators of environmental burden related to damage or importance. This is where primary fossil fuel energy used in delivering the product or service is converted into climate impacts, local air pollution, and so on. A range of eco-indicator and related environmental impact factors have been developed for use in LCA. However, ISO 14040 acknowledges that these must not be blindly applied to different temporal, spatial and product or service conditions. Hence, all results must be subject to reflective interpretation by an experienced LCA practitioner.

1.3 LCA and environmental management

LCA has considerable data requirements, and the 'question' – goal and scope – must be carefully framed. Indeed, LCA uptake has arguably been compromised by these difficulties.

Nevertheless, LCA has rapidly developed an important niche within the growing arenas of environmental management, policy and planning.

There is a spectrum of environmental management tools and techniques, ranging from the overarching 'visioning' type, to the specific 'assessment' type, to communication and reporting. Among the first type, which fosters sustainability within public, corporate and other organisations, *The Natural Step* is designed to assist an organisation to set environmental objectives and to re-think and change around these objectives. It advocates backcasting from principles and consensus processes to advance society towards sustainability through organisational change. Notwithstanding the institutional limits of such techniques (Sandström 2005), LCA can help inform change by providing information about environmental burdens of products and services associated with any organisation. LCA is particularly useful in decisions requiring comparison of environmental outcomes and can be extended through tools such as Multi-Criteria Assessment, where quantitative and qualitative information is ranked and assessed across different environmental criteria.

Systematic tools to assess, monitor, document, manage and maintain environmental performance are often modelled on ISO 14001 or similar environmental management systems, which in turn have their origins in quality management. These management approaches may also incorporate or usefully draw upon LCA, especially where specific LCA studies have been undertaken to investigate particular processes, products or services associated with the organisation concerned.

Environmental management outcomes are typically reported in order to demonstrate compliance or performance. A range of environmental reporting systems and initiatives exist, either for Global Reporting Initiative (GRI) compliance or more general 'Triple Bottom Line' or environmental corporate reports. Here, LCA mainly provides background evidence for assessment of environmental benefits, burdens or burdens foregone.

There are a wide range of other applications of LCA to 'environmental assessment'. For example, popular 'eco-footprint' and related calculators and tools may use LCA data, and advisory program information may draw on LCA results; for instance, in stating that compact fluorescent lamps generate lower environmental impacts than incandescent varieties. The 'new' tools of community engagement and 'behaviour change' for environmental outcomes may also draw on LCA data either in modelling or substantiating potential benefits of particular changes. LCA, like any modelling technique, is only as good as the modeller and the assumptions and data employed in the exercise. Indeed, this assertion is a central theme in this book.

1.4   Principles, practice and prospects for LCA: a reader's guide

While data challenges and complexities of application may have previously held LCA back, many of these are now reduced or at least are better understood. Various options exist for further improving data quality and convenience of use through quicker, easier and more ubiquitous access to LCA results while maintaining sufficient quality, accuracy and rigour. The data challenge and the balance between quality and quantity are also important themes for this book.

Given that a range of more or less qualitative judgements and unforeseen outcomes may affect the accuracy of predictions, how can LCA provide confidence in results? This is not straightforward, and indeed, LCA results have not always been accepted uncritically – often with good reason. The main strategies adopted to provide confidence are transparency and peer review. Both are strongly advocated in ISO standards.

In Chapters 2–5 LCA practice and a range of connected issues is described and critiqued. Chapter 2 charts the development of LCA and associated institutions, including SETAC, ISO, the Australian LCA Society (ALCAS) and the UNEP/SETAC Life Cycle Initiative. Policies and initiatives discussed include eco-design, cleaner production, waste management, biofuels, voluntary covenants and built environment regulations. Chapter 3 provides a detailed commentary of LCA in practice, including the different approaches and examples of how different stakeholders apply it. Chapter 4 provides a systemic critique of LCA and scrutinises its claim as an analytical approach to assessing sustainability. This chapter examines the process of problem-definition and focus on functionality and reveals LCA's limitations in dealing with wider social and consumption issues. Chapter 5 presents an empirical, practice-based perspective on how LCA varies across different geographic and environmental settings, using the example of Australia to reveal the limits of applying 'generic' data to assessing changes in 'unique' environments.

The central section of the book is dedicated to the practice of LCA, and is arranged around five principal topics: waste management, the built environment, water management, agricultural systems, and carbon and other greenhouse gas assessment. Shelter, water and food sit high on Maslow's hierarchy of essential human needs; hence the focus of three of these chapters. Waste and fossil carbon are current threats to the provision of such basic needs and raise deep questions about the efficiency and appropriateness of social and technical systems.

Chapter 6 explores waste 'management' and the role of LCA through case studies that examine the relative environmental benefits and costs of recycling paper and packaging waste, plastics and other waste fractions.

Chapter 7 commences with the policy context for 'sustainable' built environments. A series of case studies then explores the relevant applications of LCA, leading to a consideration of future directions for LCA in the built environment.

Chapter 8 focuses on water systems, including the water-related environmental impacts of technologies and product systems, and life cycle environmental impacts of water supply and servicing systems. The complex challenges of 'water in systems' and 'water service synthetic systems' leads to a wider discussion of water, design and social context.

Chapter 9 examines the LCA of agricultural systems, revealing problems with its application due essentially to the heterogeneity of such systems. Nevertheless, LCA has been successfully applied, and case studies reported here reveal results that may be regarded as controversial, counter-intuitive and/or simply surprising. Related debates and concepts, such as 'food miles', are also considered, as is the future of farming in an environmentally constrained world.

Chapter 10 examines the relationship between carbon assessment standards and LCA standards, and provides a critique of carbon offsetting from a life cycle perspective. Current and future issues for carbon management are also discussed, along with scrutiny of LCA and biomass technologies with particular reference to potential greenhouse gas savings through the substitution of fossil energy technologies.

Chapter 11 describes the rise of so-called 'quick LCA' and Life Cycle Management (LCM) tools, using contemporary examples. Discussion focuses on the need for decision support tools which provide readily available LCA information, and extends to ways in which 'decision-support' can be provided more readily to organisations. An overview of LCM leads to a stakeholder assessment of the needs of 'quick' LCA tools, which are then assembled into a series of design requirements. The development and application of two such 'quick LCA' tools are then explored through case studies.

Finally, Chapter 12 reflects critically on current LCA theory and practice and develops a prospective discussion of likely future trends in LCA to 2020. Drawing together the threads from LCA development to application and integration into current business practices, the role of LCA in influencing policy and governance in the transition towards sustainability is assessed. We also consider the factors affecting the role of LCA in assisting this transition, together with a range of other 'ingredients' that shape the prospects and uptake of LCA between now and 2020.

1.5   References

Allen D, Consoli F, Davis G, Fava J and Warren J (1997) *Public Policy Applications of Life-Cycle Assessment.* Society of Environmental Toxicology and Chemistry (SETAC), Brussels.

AS/NZS ISO (1998) AS/NZS ISO 14040:1998 'Environmental management – Life cycle assessment – Principles and framework.' Published jointly by Standards Australia and Standards New Zealand, Homebush, NSW and Wellington, New Zealand.

Barnthouse L, Fava J, Humphreys K, Hunt R, Laibson L, Noesen S, Norris G, Owens J, Todd J, Vigon B, Weitz K and Young J (1997) *Life-Cycle Impact Assessment: The State of the Art.* 2nd edn. Society of Environmental Toxicology and Chemistry (SETAC), Brussels.

Baumann H and Tillman AM (2004) *The Hitch Hiker's Guide to LCA: An Orientation in Life Cycle Assessment Methodology and Application.* Studentlitteratur, Lund, Sweden.

Christiansen K (1997) *Simplifying LCA: Just a Cut?* Society of Environmental Toxicology and Chemistry (SETAC), Brussels.

Ciambrone DF (1997) *Environmental Life Cycle Analysis.* CRC Press, LLC, Florida, USA.

Consoli F, Allen D, Boustead I, Fava J, Franklin W, Jensen AA, De Oude N, Parrish R, Postlethwaite D, Quay B, Siéguin J and Vigon B (Eds) (1993) *Guidelines for Life Cycle Assessment. A Code of Practice.* Society of Environmental Toxicology and Chemistry (SETAC), Brussels.

Curran MA (1996) *Environmental Life Cycle Assessment.* McGraw-Hill Professional Publishing, New York.

Fava JA (2002) Life Cycle Initiative: A joint UNEP/SETAC partnership to advance the life-cycle economy. *International Journal of LCA* **7**(4), 196–198.

Frankl P and Rubik F (2000) *Life Cycle Assessment in Industry and Business: Adoption Patterns, Applications and Implications.* Springer, Heidelberg.

Graedel TE (1998) *Streamlined Life-Cycle Assessment.* 1st edn. Prentice Hall, Yale.

Guinee JB (2002) *Handbook on Life Cycle Assessment: Operational Guide to the ISO Standards.* (Eco-Efficiency in Industry and Science, Vol. 7). Kluwer Academic Publishers, Dordrecht, The Netherlands.

Hofstetter P (1999) *Perspectives in Life Cycle Impact Assessment: A Structured Approach to Combine Models of the Technosphere, Ecosphere and Valuesphere.* Kluwer Academic Publishers, Dordrecht, The Netherlands.

Hunkeler D, Saur K, Rebitzer G, Finkbeiner M, Schmidt WP, Jensen AA, Stranddorf H and Christiansen K (2004) *Life-Cycle Management.* Society of Environmental Toxicology and Chemistry (SETAC), Brussels.

Hunt RG, Franklin WE, Welch RO, Cross JA and Woodall AE (1974). *Resource and Environmental Profile Analysis of Nine Beverage Container Alternatives.* EPA/530/SW-91c. United States Environmental Protection Agency, Office of Solid Waste Management Programs, Washington DC, USA.

ISO (1997) EN ISO 14040:1997 'Environmental management – Life cycle assessment – Principles and framework.' International Organization for Standardization, Geneva.

ISO (2006a) ISO 14040:2006 'Environmental management – Life cycle assessment – Principles and framework.' International Organization for Standardization, Geneva.

ISO (2006b) ISO 14044:2006 'Environmental management – Life cycle assessment – Requirements and guidelines.' International Organization for Standardization, Geneva.

Kotaji S, Schuurmans A and Edwards S (2002) *Life-Cycle Impact Assessment: Striving Towards Best Practice.* Society of Environmental Toxicology and Chemistry (SETAC), Brussels.

Kotaji S, Schuurmans A and Edwards S (2003) 'Life-cycle assessment in building and construction: a state-of-the-art report.' Society of Environmental Toxicology and Chemistry (SETAC), Brussels.

Rondinelli D and Vastag G (2000) Panacea, common sense, or just a label? The value of ISO 14001 environmental management systems. *European Management Journal* **18**(5), 499–510.

Sandström J (2005) Extending the discourse in research on corporate sustainability. *International Journal of Innovation and Sustainable Development* **1**(1/2), 153–167.

Udo de Haes H (1996) *Towards a Methodology for Life Cycle Impact Assessment.* Society of Environmental Toxicology and Chemistry (SETAC), Brussels.

Udo de Haes H, Finnveden G, Goedkoop M, Hauschild M, Hertwich E, Hofstetter P, Jolliet O, Klöpffer W, Krewitt W, Lindeijer E, Müller-Wenk R, Olsen S, Pennington D, Potting J and Steen B (2002) *Life-Cycle Impact Assessment: Striving Towards Best Practice.* Society of Environmental Toxicology and Chemistry (SETAC), Brussels.

United Nations Environmental Programme (UNEP) (2000) *Towards the Global Use of Life Cycle Assessment.* United Nations Environment Programme, Paris.

Chapter 2

The development of life cycle assessment methods and applications

Karli L Verghese, Tim Grant and Ralph E Horne

Momentum is gaining towards understanding the environmental impacts of human activities – where they occur and to what degree. Life cycle assessment (LCA) has much to offer in this regard, provided there is sufficient methodological rigour, standardisation and data to enable genuinely comparative studies and reliable results to be generated. Despite an internationally agreed scientific approach enshrined in the International Organization for Standardization (ISO) standards (14040 series), there remain various routes to consider the material and energy flows across processes in a supply chain – and therefore various possible outcomes of this consideration. As LCA practice has developed through a combination of empirical and theoretical advances, so this development has influenced both contemporary LCA studies and the range of approaches used today.

Interest in the application of LCA has grown steadily since the early 1990s. From relatively few organisations interested primarily in resource extraction (e.g. BHP Billiton and Pioneer), LCA activity widened rapidly to encompass activities as diverse as building material manufacturing, furniture production, food and beverage packaging supply chains, and water utility companies. Government departments are using LCA increasingly and exploring means by which it can provide input to policy development for a range of purposes from procurement to carbon trading. Concurrently, researchers have been developing methods and datasets in an effort to refine and improve LCA quality.

As a prelude to a discussion of current LCA practice and future prospects, this chapter critically reviews the development of LCA methods and applications. While there is specific focus on Australia, international developments in LCA are also considered as an essential backdrop, particularly regarding methodology and data development and standardisation. This chapter consists of four main parts, as follows:

- discussion of the broad policy context relating to LCA in Australia (directly and indirectly) across various sectors and jurisdictions;
- critical review of relevant LCA stakeholder forums in Australia;
- discussion of relevant regional and international initiatives; and
- an account of current challenges, developments and initiatives in Australia.

2.1 Policy context relating to LCA in Australia

Policy is invariably a reaction to an identified social, cultural, economic or environmental need. Often this identification is made by non-governmental organisations or individuals, well in advance of any policy response. International developments such as the Bruntland Commission report in 1987 and the Rio Declaration of 1992 are also broader catalysts for local, specific environmental studies or concerns which give rise to the need for LCA. The 'green' 2000 Olympic Games in Sydney responded to the growing environmental agenda manifested in such developments and generated a step-change in LCA uptake and development in Australia.

From such examples, it is evident that LCA-specific policies are rare, but the opportunities for increased uptake of LCA as a response to wider policy developments are increasing. Moreover, the potential exists for a mutually reinforcing relationship between the growing awareness of environmental issues and the use of LCA within a particular industry sector or around a particular environmental issue. Interest in using LCA in Australia began in the early 1990s in the products sector with the publication of 'More with less – initiatives to promote sustainable consumption' (Deni Greene Consulting Services *et al.* 1996). This study examined sustainable consumption in Australia and related contemporary initiatives. It generated interest with the federal government, which instituted two industry environmental programs: one at the manufacturing level – the cleaner production case studies (Australian Government n.d.), and the second in the area of design, with the EcoReDesign program (Gertsakis *et al.* 1997), which included LCA as a core component (see Section 2.1.2).

Interest in LCA increased with the awarding of the 2000 Olympic Games to Sydney. Tender documents for infrastructure projects for the Olympics required companies to demonstrate the environmental credentials of their materials and products. This, in turn, led to government recognition of the need for LCA data relevant to Australia. State government Environment Protection Authorities (EPAs) and the federal government funded the compilation of an LCA inventory dataset in the mid 1990s: the National Life Cycle Inventory (LCI) Database (see Section 2.2.1). Then, in the late 1990s, the Victorian Government funded research to identify the relative environmental costs and benefits of recycling packaging materials compared with landfill. Subsequent studies in 2003 and 2005 examined waste recovery options and commercial, industrial and construction or demolition waste respectively (see Section 2.1.3 and Chapter 6). In 2000, The Australian Greenhouse Office (AGO) funded work to scope LCA studies specifically on energy and greenhouse gases, which led to the use of LCA in applications by industry for Greenhouse Friendly™ accreditation (see Section 2.1.4 for a definition). In the early 2000s, water utility authorities, such as Yarra Valley Water and Sydney Water, commissioned studies to investigate the benefits of different water supply and treatment services. In 2004 EPA Victoria identified the importance of LCA by including specific reference to it in their strategic plan. This subsequently led to the formation of a Sustainability Covenant between EPA Victoria and the Plastic and Chemical Industry Association (PACIA), which promoted life cycle thinking more broadly among raw material suppliers and manufacturers (see Section 2.1.7).

2.1.1   The built environment

In 1993, when the 'green' Olympic Games in 2000 were awarded, building material and product suppliers such as BHP Steel, James Hardie and Pioneer initiated studies to consider the life cycles of their products. These companies realised that they needed to understand better the environmental impacts occurring in the production of their materials so that they could demonstrate their environmental credentials and so be awarded contracts to supply materials in the construction phase.

As New South Wales was the 'home' state of the Olympics, the New South Wales Public Works Department rapidly developed an interest in the environmental performance of buildings and public infrastructure. The department's work centred upon building and construction of institutional buildings and, as the Olympic infrastructure began to develop, it focused on the Olympic stadium and associated buildings. The federal government, through Environment Australia – the Department of Environment and Heritage, funded a project undertaken by the LCA group at the Centre for Design at RMIT University, which included a critical examination of tools for undertaking LCA in the building and construction sector. The project resulted in a website, *Greening the Building Cycle: Life Cycle Assessment Tools in Building and Construction* (http://buildlca.rmit.edu.au), which was completed in 2001. The website promoted LCA as a tool to assess the environmental impacts of building materials and building systems in Australia with the aim of improving the environmental performance of the building and construction sector. It contained case studies, analyses of national and international tools, environmental performance data sheets and a decision support tool. Contemporaneously, the Australian Commonwealth Scientific and Research Organisation's (CSIRO) Division of Materials undertook research on the embodied energy of building and construction materials and processes. This was based on input-output analysis (see Chapter 3) and led to the development of input-output models at Deakin University.

2.1.2   Product design

In 1993, the federal government established the National EcoReDesign Demonstration Project, undertaken by the Centre for Design at RMIT University, which drew heavily on LCA for the re-design of products from seven Australian manufacturers. Between 1994 and 1997, the project re-designed products to demonstrate what 'design for the environment' could achieve for business success and the environment (Environment Australia 1997). The companies involved and the re-designed products were:

- Caroma – a high-performance water conservation device
- Blackmores – environmentally preferred packaging with the proviso that it would not compromise existing performance criteria such as point-of-sale appeal and ease of use
- Imaging Technologies – an office vending machine for collecting and recycling toner cartridges from fax and photocopier machines
- Email – a clothes washing machine designed for maximum performance and resource efficiency
- Southcorp Whitegoods – a dishwasher designed for recyclability with improved water and energy efficiency, and eco-based control panels
- Schiavello Commercial Interiors – modular office furniture designed for long life and recyclability, using environmentally preferred materials.

The LCA work undertaken in this project was limited by the lack of the availability of local Australian data, and this experience formed one of the key drivers in the subsequent development of the National LCI Database.

Other LCAs driven by product design have been undertaken in the commercial furniture, the packaging and the electronics sectors. Requests from suppliers for information, inclusion of environmental criteria in tendering documents, introduction of regulations and/or guidelines or wanting to improve the knowledge within the company of the environmental impacts associated with materials and processes are all factors that drive individual companies to commission LCAs.

2.1.3   Waste management

LCA has had a prominent role in waste management (see also Chapter 6). In Australia, the Victorian government led the way in 1997 by commissioning significant LCA studies into waste management. These studies guided the development of the first National Packaging Covenant in 1999. The Victorian government's Greenhouse Strategy 2002 identified waste as an important contributor to the greenhouse effect through methane emissions from landfill, transport and processing of waste, and indirectly via lost savings that could be gained through recycling of valuable materials. The 2002 Victorian Solid Waste Strategy, which was designed to deal with a broad range of waste management issues including greenhouse gas emissions, aimed to develop policies and programs required to reduce the volume of waste going to landfill and to ensure the management of waste occurs in a manner that promotes optimal environmental outcomes including the reduction of greenhouse gas emissions (The State of Victoria 2005). The strategy considered waste from a variety of sources including municipal, commercial and industrial, and construction and demolition waste. EcoRecycle Victoria then commissioned two further LCAs: the first to evaluate the environmental impacts of a range of waste management scenarios, and the second to examine the environmental benefits of recycling construction and demolition, and commercial and industrial waste, such as concrete, timber, bricks, used tyres and batteries. Each of these major studies allowed for the quantification of the benefits and impacts of recycling waste materials, and they have been used to guide policy development in addition to being used as a communication tool.

2.1.4   Greenhouse issues

The Commonwealth Government's 1998 National Greenhouse Strategy, Measure 4.17 'Life cycle energy analysis', states:

> Life cycle energy issues will be pursued through the following actions (a) governments, in consultation with industry, will develop a database and nationally accepted methodology for life cycle energy analysis and (b) based on these life cycle analyses, policies/programs will be developed and implemented to encourage producer responsibility for sourcing of materials, product design and manufacture, product operating efficiencies and product disposal, as a means of improving greenhouse outcomes (AGO 1998, pp. 50–1).

In realising the strategy, a scoping study was undertaken to establish what a life cycle energy database and methodology might look like (Lundie *et al.* 2001). The study recommended that the AGO should concentrate its efforts on having full LCAs undertaken in the Australian context to support the development of data quality criteria and the initiation and implementation of a nationally accepted LCA database. The project stalled amid valid concerns about creating a life cycle database that only considered greenhouse and energy issues and not a fuller range of life cycle issues. In 2003, the AGO developed the Greenhouse Friendly™ program that certifies products and services as being 'carbon neutral'. The method involves undertaking a greenhouse gas emission LCA of a product or service and then offsetting the greenhouse emissions with certified greenhouse gas abatement options (see Chapter 10 for a further examination).

2.1.5   Biofuels

In 1999, then Australian Prime Minister Mr Howard stated that the AGO may certify additional fuels to diesel as being eligible under the Diesel and Alternative Fuels Grants Scheme. This prompted the AGO to commission an LCA by CSIRO and research partners to compare road transport fuel emissions, including but not limited to greenhouse gas emissions, to determine which fuels were appropriate to be certified under the scheme. A scoping study was conducted in 2000 (Beer *et al.* 2000). This study reviewed earlier studies conducted in the United States of America (USA) and Europe on fuels including biodiesel. The scoping study led to a full fuel-cycle analysis of alternative fuels to compare their emissions of greenhouse gases and other air toxins (Beer *et al.* 2001). The fuel against which other fuels were compared was low sulphur diesel, and the fuels examined were:

- diesel fuels
- biodiesel and canola oil – the biodiesel came from five upstream sources: canola, soy and rape crops, tallow and waste cooking oil

- gaseous fuels
- hydrated ethanol-based fuels, namely Diesohol, which is a blend of 15% ethanol with low sulphur diesel and an emulsifier, and hydrated ethanol produced from seven upstream sources
- hydrogen
- light vehicle fuels, namely premium unleaded petrol (PULP) blended with 10% anhydrous ethanol, and anhydrous ethanol blended with 15% PULP – again, the ethanol was produced from seven upstream sources.

The LCA concluded that, in general, biodiesel and ethanol emit less greenhouse gases than diesel, taking into account both the exbodied emissions of the fuels themselves and upstream activities. The same was found to be true for most but not all other air toxins examined (see also Chapter 10 for further discussion on biofuels).

Following this report, the federal government announced that 350 megalitres of biofuels would be included in the national fuel mix by 2010 (Department of Industry Tourism and Resources 2007). In 2003, CSIRO, the Australian Bureau of Agriculture and Resource Economics (ABARE) and the Bureau of Transport and Regional Economics (BTRE) conducted a desktop study of the appropriateness of the target (CSIRO *et al.* 2003). This study concluded that the cost of government support for biofuels was likely to outweigh the benefits. In response, the Prime Minister established a Biofuels Taskforce to review policy on biofuels, which reported that Australia was unlikely to meet the 350-megalitre target by 2010, based on the current state of the industry and low consumer demand. The Taskforce also found that the potential reductions in greenhouse gas emissions from biofuels were not sufficient to warrant a significant policy change in their favour (Department of the Prime Minister and Cabinet 2006). The Prime Minister then reconfirmed the target and announced the development of industry action plans to achieve it (Department of Industry Tourism and Resources 2007).

2.1.6 Water management

LCA has increasingly focused on water and wastewater management since the late 1990s. From 2000, Sydney Water undertook LCAs on specific issues including biosolids treatment and alternative (reticulated and non-reticulated) water delivery options (Sydney Water 2000). Yarra Valley Water in Victoria has also undertaken LCAs on alternative water and sewage servicing options in a series of LCAs from 2003. (See Chapter 8 for an examination of Yarra Valley Water's LCA work and other water management issues in LCA.)

2.1.7 Sustainability covenants

Sustainability covenants, incorporated into the Victorian *Environmental Protection Act 1970*, are voluntary agreements through which regulators and organisations or groups of organisations can explore ways to reduce environmental impacts and increase the resource efficiency of their products and services (EPA Victoria 2007). This policy initiative marks a co-regulatory partnership approach between the regulator and commercial organisations. The intent is to provide a more efficient means of maintaining higher environmental standards than the previous 'command and control' approach, which was principally based on works approvals and licences. Hence, sustainability covenants are seen as an instrument to broaden business focus to include social and environmental considerations compared with traditional point source management.

As mentioned above, EPA Victoria and PACIA signed a voluntary sustainability covenant in January 2004. One of the commitments of the covenant is to develop:

> guidance materials and decision support tools (including, for example, life cycle assessment and environmental management accounting) that will assist the industry to measure resource efficiency and potential ecological impact across the

life cycle of products; and identify opportunities to increase resource efficiency and reduce ecological impacts throughout the life cycle of products. (EPA Victoria and PACIA 2004, p. 7)

2.2   LCA stakeholder forums in Australia

LCA is a new and relatively data-intensive technique, with information requirements that invariably extend beyond traditional or existing organisational and database boundaries. Therefore, its development requires the setting up and successful maintenance of relationships of trust and mutual interest between organisations that either have no previous relationship or have very different relationship histories. Cognisant of this, the LCA research community in Australia has attempted to build appropriate forums through which these new relationships can be created and developed. The forums fall into four main categories, each of which is examined below:

- project-based partnerships
- round-table forums
- a specific society, the Australian Life Cycle Assessment Society (ALCAS)
- conferences.

2.2.1   Life Cycle Inventory development

The national LCI project is a key example of an attempt to create new partnerships for LCA development around a project format, with mixed success. The project was originally commissioned by environment departments from both the federal and four state governments (Victoria, New South Wales, Queensland and South Australia) in 1997. The research was undertaken through a collaboration of the Centre for Design at RMIT University and the Centre for Water and Waste Technology (CWWT) at the University of New South Wales through the Co-operative Research Centre (CRC) for Waste Management and Pollution Control (CRC WMPC). The project focused on building and packaging materials, transport and energy. Earlier attempts to develop public data projects jointly with industry had not been particularly successful (e.g. International Iron and Steel Institute's LCI on steel in 1996 (IISI n.d.) and Cement and Concrete Association of Australia's LCI on concrete (CandCAA n.d.), which were not consistent with each other in method). However, the national LCI project drew heavily on a mix of public data sources in Australia in combination with relevant international LCA data. As LCI elements were developed, industry review in Australia was invited. Responses were limited and relatively few immediate updates were made to the datasets. Subsequently, these have been regularly updated as new information becomes available (e.g. through the National Pollutant Inventory and the National Greenhouse Gas Inventory).

Since the national LCI project was initiated, data resources in Australia have improved through a range of publicly funded LCA projects and industry-based initiatives. In the public arena, LCA work has been undertaken on paper and packaging waste management, residual waste management, organic waste management, alternative fuels for heavy vehicles and management of waste oils. Industry-based initiatives include inventories developed for vinyls, steel, aluminium, cement and concrete. Today, LCA data relevant to Australia is available for a wide range of areas including waste management (landfill, material recycling and waste-to-energy), transport and fuels, packaging, buildings, raw materials, food and agriculture, process engineering, product design and life cycle impact assessment.

The Centre for Design at RMIT University has been the de facto clearing house for LCI data storage, although ALCAS is increasingly developing the capacity to fulfil this role. In 2006,

ALCAS and CSIRO Sustainable Ecosystems launched an Australian LCA database initiative (AusLCI) (CSIRO Sustainable Ecosystems 2006b) as discussed further below.

2.2.2   Roundtable forums and the Australian Life Cycle Assessment Society (ALCAS)

The Centre for Water and Waste Technology at the University of New South Wales initiated a series of LCA roundtable forums in 1996. These were intended to provide a conduit for multi-stakeholder participation in LCA, involving industry, government, academic institutions and other LCA practitioners. Initially, several roundtables were held each year, at which participants had the opportunity to inform others of projects that were underway and present interim and final results, discuss developments and research needs, and reflect on critical issues in LCA development.

ALCAS grew out of the relatively informal roundtable forums after participants decided to formalise its structure and membership (ALCAS 2007). ALCAS is now a professional organisation for people interested in the practice, use, development and interpretation of LCA. It is not-for-profit and has individual and corporate members principally from industry, government, consulting and academic institutions. The purpose of the society is two-fold: to promote and foster the appropriate development and application of LCA methodology in Australia and internationally with a view to making a positive contribution to ecological sustainable development (ESD); and to represent the Australian LCA community in the international arena. The specific objectives of ALCAS are to:

- promote and foster the appropriate application of LCA in Australia
- promote and foster the responsible development of LCA methodology in Australia with consideration of international initiatives and commensurate with local conditions
- foster links with the international LCA community
- organise a regular LCA roundtable to facilitate information exchange and discussion on LCA among stakeholder groups
- contribute to national policies, positions and approaches on LCA and its applications both nationally and internationally
- increase education and awareness of LCA among stakeholders including industry, academic institutions, government, non-governmental organisations, LCA practitioners, end users and the general public
- develop a national competence in LCA to meet the environmental challenges both locally and internationally.

2.2.3   National LCA conferences (1996–2006)

The first national conference on LCA, held in Melbourne in 1996, highlighted the need for public data on LCA in Australia. This in part led to the initial national LCI project. The second national conference was held in Melbourne in February 2000, and the focus shifted from data issues (in 1996) to impact assessment and characterisation of the Australian environment. Since 2000, conferences have been held approximately biennially, with ALCAS taking an increasingly significant role in communicating LCA to a wider audience. The result has been significant growth in delegate numbers. The third national conference was held on the Gold Coast in Queensland, in July 2002, and the fourth in Sydney in February 2005. The fifth, in Melbourne in November 2006, had sufficient presenters and delegates to warrant a multi-stream format, with significant interest from organisations with little previous connection with ALCAS or LCA.

2.3 Regional and international initiatives

2.3.1 Development of LCA through the Society of Environmental Toxicology and Chemistry and the International Organization for Standardization standards

It is impossible to discuss the evolution of LCA without considering the work of the Society of Environmental Toxicology and Chemistry (SETAC). The involvement of SETAC has had such an enduring effect on the pathways and scope of LCA that environmental toxicology has occupied a significant proportion of scientific efforts in LCA. Development of toxicological models and extensive databases on chemical impacts has culminated in the Omnitox project (Molander *et al.* 2004). SETAC's contribution to forming LCA method began with SETAC guidelines to LCA in 1992 (Fava *et al.* 1991; Fava *et al.* 1992). The disadvantage of SETAC's work is a bias towards European and North American development (particularly the former). SETAC in the Asia Pacific region has never had a strong LCA focus, so the SETAC developments have focused on issues more relevant to the highly populated areas of Europe, mostly relating to increases in concentrations of pollutants. Elsewhere in the world, and particularly in Australia, the issues of land use and water resources are critical and have generally been paid less attention in the LCA debate.

The development of documents and working groups in SETAC have shaped much of the LCA content in the international standards process undertaken by ISO, which began just as LCA was being formalised by SETAC. The ISO process took what SETAC had initially developed, and over four years developed an international consensus that became the ISO Standards on LCA (ISO 14040 series). For better or worse, the standards had to be broad enough to incorporate the diversity of LCA users and the lack of consensus on specific methodological issues in LCA. The issue of impact assessment was one of the most difficult standards to complete, with discursive debate on the use of indicators and weighting.

As the long-awaited LCA standards were nearing completion, a synthesis project was proposed to bring the stages of LCA into two overarching standards: the framework and requirements for LCA: ISO 14040:2006 (ISO 2006a), *Environmental management – Life cycle assessment – Principles and framework*; and ISO 14044:2006 (ISO 2006b), *Environmental management – Life cycle assessment – Requirements and guidelines* (Finkbeiner *et al.* 2006). Significant developments in LCA that are not in the standards include how to undertake weighting in LCA (which is not condoned for public comparative studies), and a recognition of the distinction between consequential and attributional approaches.

2.3.2 Developments through the United Nations Environment Programme/Society of Environmental Toxicology and Chemistry Life Cycle Initiative

With the completion of the ISO standards, it was clear that standards on their own would not promote the use and development of LCA and that the breadth of the standards still allowed significant diversity in LCA practice, making it difficult for newcomers to navigate appropriate methods. SETAC approached the United Nations Environment Programme (UNEP) to work together on building international consensus on LCA use and development. UNEP were most interested in how emerging economies might be included in the LCA debate and development. From these two differing interests, the United Nations Environment Programme/Society of Environmental Toxicology and Chemistry (UNEP/SETAC) Life Cycle Initiative (LCI) was born (UNEP and SETAC n.d.a). Stage one developed extensive working groups in three broad areas (Udo de Haes *et al.* 2002):

- application of and education on LCA and life cycle thinking
- development and enhancement of sound LCI data and methods
- development and enhancement of sound life cycle impact assessment data and methods.

Stage 1 of the LCI was completed after three years with mixed results, as consensus on methodology was again difficult to achieve, although through the LCI and other developments, regional LCA networks were formed in Africa and Latin America (UNEP and SETAC n.d.b).

2.3.3   LCA activity in the Asia Pacific region

In the Asia Pacific region, Japan has for many years led development both at a national level and in fostering LCA application in Asia and Asian Pacific countries. A series of workshops sponsored by the Asia Pacific Economic Cooperation (APEC) were held from 1998 to 2006 and are continuing. The focus of these workshops has been on capacity building, inventory development, impact assessment and specific applications of LCA in waste management and agriculture. Other projects in the region include a project by the Asia Productivity Organisation on LCA development, and a Japanese project to collect basic data on electricity and materials from Asian Pacific countries. This has helped to foster growth in LCA in the region, particularly in Thailand and Malaysia.

2.4   Current challenges, developments and initiatives in Australia

One of the key challenges for LCA in Australia is the ongoing development of LCIs. This is discussed prior to drawing more general conclusions.

2.4.1   LCI in Australia

In 2006, the National Life Cycle Inventory Database (AusLCI) project led by ALCAS and CSIRO Sustainable Ecosystems in conjunction with EPA Victoria, PACIA, Sustainability Victoria, Centre for Design at RMIT University, and the Forest and Wood Products Research and Development Corporation was initiated with the aim of developing national inventory data in cooperation with industry (CSIRO Sustainable Ecosystems 2006b). The initiative aims to set up a framework and processes for the collection and publishing of consistent LCA data in Australia through involvement of industry, research organisations and government. This is a long-term project, and although it has provided new data developments, it is likely to take several further years for a consistent dataset to be developed. The project is similar in design – and includes plans to integrate where possible with – other national data projects in Switzerland, the USA, Malaysia and Thailand, and a European-wide project. The AusLCI initiative activities are also designed to promote the development of life cycle impact assessment methods in Australia.

The initiative is a cross-sectoral project to develop data in many sectors; including building materials, energy systems, agriculture, packaging and waste management (Table 2.1). Each sector and companies within the sector are responsible for participating in and funding the data collection, with support from the initiative to review and publish, and develop agreed guidelines and methods for undertaking the LCI (CSIRO Sustainable Ecosystems 2006a).

Table 2.2 presents an overview of the status of LCI in Australia, identifying sectors and/or industries where:

- research on data collection has been undertaken for more than five years, or
- there is increasing interest in LCA (i.e. work has been conducted in the past five years), or
- there is still a lack of interest in data collection.

As discussed earlier in this chapter, public LCI work has been underway in Australia since the early 1990s. However, there has been relatively little coordination between different

Table 2.1   Life cycle initiative data sectors

Life cycle stage	Sector
Raw material extraction	Basic minerals: bauxite, iron ore, ilmenite, copper zinc ores, gravel, sand, soil, limestone, gypsum, forestry, uranium
Fuels	Coal, oil and gas, refinery operation, biofuels, biomass
Processing and converting	Basic materials including (but not limited to): slab steel, aluminium ingot, polymers (PE, PP, PVC, PA, PS, ABS), rubbers (SBR), wood processing
Manufacturing	Metal forming, plastics molding, drilling
Transport	Road, rail, sea, air, freight and passenger services calculated using mass, kilometre, volume kilometre
Energy and heat	Energy from fuels, electricity supply, machinery operation
Use	
Waste management	Landfill, recycling, composting, waste-to-energy

ABS (Acrylonitrile butadiene styrene), PA (polyamid), PE (polyethylene), PP (polypropylene), PVC (polyvinyl chloride), PS (polystyrene), SBR (Styrene butadiene rubber)

development projects and no consistent method for each project. A review of data development is provided here to indicate various activities, and information and knowledge development in each sector.

2.4.1.1   Metals

The steel and aluminium industries in Australia have both been involved in LCA data development over the last 10 years, although neither has publicly released inventory data in a systematic way. Instead, limited inventory data has been provided on a case-by-case basis to research projects. Basic metal production data and some finished product data, such as sheet production, have been provided in some areas. For non-ferrous metals other than aluminium, no specific company data has been collated. The University of New South Wales/RMIT inventory project developed steel data from public information made available by BHP, and the gaps were filled from inventories developed in the USA. Aluminium data in this same project was developed principally with data from the Australian Aluminium Council. Other LCA work has derived steel data from a European inventory (Boustead model), and CSIRO Minerals has undertaken LCA for copper and nickel production based on different production routes. Data for this project was based on theory rather than on 'average Australian consumption'. While there is scope to update and enrich steel and aluminium data, there is also a particular need to develop LCI data for other metal sectors.

Table 2.2   Status of life cycle initiative in Australia

Industry is mature with LCI	Industry is interested	Industry has low interest
Steel	Timber	Electricity
Concrete	Plastics	Transport
Vinyl	Water	Imports (e.g. electronics)
Aluminium	Waste management	
	Agriculture	
	Chemicals	
	Automotive	
	Packaging	

2.4.1.2   Minerals and glass

Data on non-metalliferous minerals is largely limited to studies by CSIRO Minerals and the University of Sydney, although the focus of these has not been public LCI data. Hence, there is considerable scope to improve data provision for LCA; for example, for lime, gypsum and soda ash – all important components of building materials. Pilkington (Australia) (a flat glass producer) has undertaken some LCI development in Australia while more detailed work has been undertaken by the parent company in the United Kingdom. Bottle glass data was developed as part of the original LCI project in 1998 by University of New South Wales/RMIT, with some input from ACI Glass Packaging.

2.4.1.3   Building products

Brick and tiles industry associations have undertaken LCA studies with the University of Newcastle and have provided this data for specific projects. The Concrete Industry Association has developed inventories for Australian cement and concrete – not publicly released, but released for specific projects. The Cement Industry Federation has also produced data relating to energy and materials used in cement production. Also, Independent Cement and Lime (ICL) has undertaken some LCA work on its blended cement products incorporating blast furnace slag. Older data on concrete and cement production was established in the original 1998 University of New South Wales/RMIT LCI project. The LCA Design project (developed by the CRC for Construction and Innovation) has developed cement and concrete data for individual states. For timber, hardwood and softwood data was developed as part of the 1998 LCI project by University of New South Wales/RMIT, with some input from the University of Tasmania, and an industry-wide LCI project has been undertaken (2007–08).

2.4.1.4   Liquid and gaseous fuels

There has been significant industry activity in LCI development related to liquid and gaseous fuels. For example, BP has produced an LCA for Greenhouse Friendly™ certification, and other producers have been or are participating in LCA projects. These are unpublished and often originally government-supported studies, then followed up by industry-funded work to support, refine or refute the original studies. Much of the LCA emphasis on fuels production has focused on greenhouse gas emissions, so additional work is required to cover other emissions and environmental impacts.

2.5   Conclusions

From piecemeal origins, LCA has matured rapidly as a technique in Australia. Clearly, data remains a key challenge for LCA development, and the efforts and issues summarised above indicate the need for the development of a more comprehensive National LCI Database. In parallel, the 'professionalisation' of LCA is an increasingly widespread phenomenon. For example, ALCAS now has the status of a not-for-profit professional association. Such developments are driving the maturation of LCA practice and uptake. In relation to theoretical and methodological development, there remains a tension between the need for standardisation of approaches in line with ISO standards (and beyond) and the competing need for flexibility in approach. The former aids the ability to compare results across studies, while the latter helps maintain a critical and experimental focus that is essential in research, innovation and development.

Against the backdrop of LCA maturation is the ongoing need to link LCA practice in Australia with international developments. The importance of initiatives to create international consistency sits alongside national or more localised projects, data development and research. A major task that remains largely incomplete in Australia is the development of a set of country-specific eco-indicators or impact assessment algorithms. Further discussion of the future prospects and development needs for LCA in Australia is provided in Chapter 12.

2.6   References

AGO (1998) 'The national greenhouse strategy – strategic framework for advancing Australia's greenhouse response.' Australian Greenhouse Office. Commonwealth of Australia, Canberra.

ALCAS (2007) *About ALCAS, Australian Life Cycle Assessment Society, Melbourne.* Retrieved 27 October 2007 from <http://www.alcas.asn.au>.

Australian Government (n.d.) *Design for the Environment, Department of the Environment, Water, Heritage and the Arts, Canberra.* Retrieved 11 February 2008 from <http://www.environment.gov.au/settlements/industry/corporate/dfe.html>.

Beer T, Grant T, Brown R, Edwards J, Nelson P, Watson H and Williams D (2000) 'Life-cycle emissions analysis of alternative fuels for heavy vehicles – Stage 1.' CSIRO Atmospheric Research Report C/0411/1.1/F2 to the Australian Greenhouse Office, Melbourne.

Beer T, Grant T, Morgan G, Lapszewics J, Anyon P, Edwards J, Nelson P, Watson H and Williams D (2001) 'Comparison of transport fuels. Final report EV45A/2/F3C to the Australian Greenhouse Office on the Stage 2 study of life-cycle emissions analysis of alternative fuels for heavy vehicles.' CSIRO Atmospheric Research, Aspendale, Victoria.

CandCAA (n.d.) *Life Cycle Assessment of Buildings in Australia – Case Studies, Cement and Concrete Association of Australia.* Retrieved 1 November 2007 from <http://www.concrete.net.au/Life_cycle/case_studies/lcadata.html>.

CSIRO, BTRE and ABARE (2003) 'Appropriateness of a 350 ML biofuels target. Report to the Australian Government. Department of Industry, Tourism and Resources, Canberra.' Commonwealth Scientific Industry Research Organisation (CSIRO), Bureau of Transport and Regional Economics (BTRE) and Australian Bureau of Agriculture and Resource Economics (ABARE).

CSIRO Sustainable Ecosystems (2006a) *The Australian Life Cycle Inventory Database, CSIRO Sustainable Ecosystems, Melbourne.* Retrieved 11 February 2008 from <http://www.auslci.com/>.

CSIRO Sustainable Ecosystems (2006b) *Sustainable Products, Sustainable Futures.* Media Release 06/240, CSIRO, Melbourne. Retrieved 27 October 2007 from <http://www.csiro.au/news/ps2i8.html>.

Deni Greene Consulting Services, Australian Consumers Association and National Key Centre for Design at RMIT University (1996) 'More with less – initiatives to promote sustainable consumption'. Environmental Economics Research Paper No.3, for the Department of the Environment, Sport and Territories. Commonwealth of Australia, Canberra.

Department of Industry Tourism and Resources (2007) *Alternative Transport Fuels: Biofuels.* Retrieved 1 November 2007 from <http://www.industry.gov.au/content/itrinternet/cmscontent.cfm?objectID=5A4C427F-DF31-42CC-110C49161D4B3D51>.

Department of the Prime Minister and Cabinet (2006) Australian Government Biofuels Taskforce, Australian Government, Canberra. Retrieved 11 February 2008 from <http://www.pmc.gov.au/taskforce>.

Environment Australia (1997) *The EcoReDesign Project Fact Sheet.* Commonwealth of Australia, Canberra. Retrieved 24 October 2007 from <http://www.deh.gov.au/settlements/industry/corporate/eecp/publications/fs-ecoredesign.html>.

EPA Victoria (2007) *Sustainability Covenants, EPA Victoria, Melbourne.* Retrieved 27 October 2007 from <http://www.epa.vic.gov.au/bus/sustainability_covenants/default.asp>.

EPA Victoria and PACIA (2004) *Sustainability Covenant between Environment Protection Authority and the Plastics and Chemicals Industry Association Inc, EPA Victoria, Melbourne.* Retrieved 27 October 2007 from <http://www.epa.vic.gov.au/bus/sustainability_covenants/docs/PACIA_covenant.pdf>.

Fava J, Denison R, Jones B, Curran MA, Vigon B, Selke S and Barnum J (1991) 'A technical framework for life cycle assessment.' Society of Environmental Toxicology and Chemistry (SETAC) and the SETAC Foundation for Environmental Education: Pensacola, Florida, USA.

Fava J, Jensen AA, Lindfors L, Pomper S, De Smet B, Warren J and Vigon B (1992) 'Life cycle assessment data quality: a conceptual framework.' Society of Environmental Toxicology and Chemistry (SETAC) and the SETAC Foundation for Environmental Education: Pensacola, Florida, USA.

Finkbeiner M, Inaba A, Tan R, Christiansen K and Klüppe H-J (2006) The new international standards for life cycle assessment: ISO 14040 and ISO 14044. *International Journal of Life Cycle Assessment* 11(2), 80–85.

Gertsakis J, Lewis H and Ryan C (1997) 'A guide to EcoReDesign.' National Centre for Design at RMIT University, Melbourne.

IISI (International Iron and Steel Institute) (n.d.) The life cycle of steel (LCA/LCI). Retrieved 2 November 2007 from <http://www.worldsteel.org/?action=storypagesandid=233>.

ISO (2006a) ISO 14040:2006 'Environmental management – Life cycle assessment – Principles and framework.' International Organization for Standardization, Geneva.

ISO (2006b) ISO 14044:2006 'Environmental management – Life cycle assessment – Requirements and guidelines.' International Organization for Standardization, Geneva.

Lundie S, James K, Sonneveld K and Grant T (2001) 'State of life cycle energy analysis in Australia'. Report to the Australian Greenhouse Office. Centre for Water and Waste Technology at the University of New South Wales, the Centre for Packaging, Transportation and Storage at Victoria University and the Centre for Design at RMIT University, Sydney and Melbourne.

Molander S, Lidholm P, Schowanek D, del Mar Recasens M, Fullana i Palmer P, Christensen F, Guinée J, Hauschild M, Jolliet O, Carlson R, Pennington D and Bachmann T (2004) Preamble: OMNIITOX – operational life-cycle impact assessment models and information tools for practitioners. *International Journal of Life Cycle Assessment* 9(5), 282–288.

The State of Victoria (2005) 'Sustainability in action: towards zero waste strategy.' EcoRecycle Victoria, Melbourne.

Sydney Water (2000) 'Sydney Water's 2000–2005 environment plan.' Sydney Water, Sydney.

Udo de Haes H, Jolliet O, Norris G and Saur K (2002) UNEP/SETAC life cycle initiative: background, aims and scope. *International Journal of Life Cycle Assessment* 7(4), 192–195.

UNEP and SETAC (n.d.a) *The Life Cycle Initiative.* United Nations Environment Programme (UNEP) and the Society of Environmental Toxicology and Chemistry (SETAC), France. Retrieved 11 February 2008 from <http://lcinitiative.unep.fr/>.

UNEP and SETAC (n.d.b) *Regional Networks – The Life Cycle Initiative.* United Nations Environment Programme (UNEP) and the Society of Environmental Toxicology and Chemistry (SETAC), France. Retrieved 11 February 2008 from <http://frl.estis.net/builder/includes/page.asp?site=lcinitandpage_id=1E7C0012-A731-4136-907E-373473305B84>.

Chapter 3

Life cycle assessment in practice

Tim Grant

This chapter introduces and critically evaluates methods and approaches in LCA practice, both in Australia and around the world. Generic reference to 'life cycle assessment' (LCA) includes a wide range of practices and approaches, each with strengths and weaknesses and offering a range of appropriateness to different settings and contexts. Cutting across this diversity, however, runs a core set of elements that are common and necessary to all successful LCA practice. These common elements are discussed, following an introduction to the diversity of approaches used.

3.1 Introduction: a typology of practices

LCA is often referred to as a single technique. However, within the general principles of adding up environmental impacts along a supply chain and representing these as environmental indicators, there is a plethora of approaches and scales within which LCA is undertaken. Variation in LCA practice is a result of differences in questions, availability of resources and data, and variations in environmental or socioeconomic conditions. Added to this is a wide range of approaches allowed by LCA standards, and significant differences across practitioners' preferences, so it is unsurprising that there is considerable diversity in the way studies are conducted.

The description of the problem or question is the appropriate starting point in all LCA studies. However, from this initial point, diversity in practice starts to creep in. The question can be vague. For example: 'Where are the environmental impacts in my supply chain?' Or very specific; for example: 'Are plastic bumper bars superior from a greenhouse- and resource-depletion perspective to steel bumper bars in Australian manufactured cars?' The questions in LCA can evolve over the study, and the extent to which resources are committed before the first insights and answers are obtained is often the extent to which resources are wasted on the wrong question. Efficient and agile use of LCA dictates an iterative approach which feeds back early results of LCA studies into the later stages of data collection, reporting, impact assessment and interpretation.

Diversity in practice then permeates the remainder of the LCA process, starting with the establishment of the assessment's scope. Three major variables in the scope are typically adjusted according to the type of assessment being undertaken. They are:

- the number of life cycle stages
- the number of environmental impacts or indicators to be considered
- the quantity of data locally collected specifically for the LCA study.

There is also an area of LCA practice where the scope is expanded beyond the traditional functional unit approach to include social and/or economic indicators. There is no clear boundary or point at which such studies cease to be called LCA and become part of 'integrated assessments'.

Two other distinctions need to be made in describing LCA practice. The first is between 'bottom-up' process analysis studies (see Section 3.3) and 'top-down' input-output studies (and a continuum of hybrid approaches in-between). The second distinction is between consequential and attributional approaches to LCA modelling, which is similar to the distinction in economics between average and marginal costing. Attributional LCA focuses on development of average impacts for a supply of commodities and production processes and is historically how most LCAs have been conducted. Consequential LCA considers the impacts of decisions (i.e. the net environmental impact from marginal changes in production and consumption). It takes account of issues of scale, location and market conditions to predict which parts of supply chains will increase or decrease with changes in demand (Ekval and Weidema 2004).

3.2   The generic elements of LCA

Within the broad category of LCA practice, certain key elements, are always required, and these are described in this section.

3.2.1   Functional unit

The basis for LCA is the calculation of environmental impacts for the delivery of specific functions or utilities. Conventionally, the functional unit should be the same for all options analysed. The defining of appropriate functional units is a challenge in LCA primarily because options may have secondary functions that need to be balanced. The concept of the functional unit has been expanded by some practitioners to include the economic values of the options analysed.

Conceptually, the functional unit is defined so that results from the LCA can be used to promote a substitution of the option(s) with lower environmental load for the option(s) with greater environmental load. With this in mind, the functions need to be so close to equal that a 'rational economic person' (or perhaps a 'rational environmental person') would consider the utility or benefit of each option to be similar. In this way, for example, if different vehicle options were manufactured using different materials, most people would not consider this a barrier to purchase. However, if the vehicle options differed in their durability, efficiency, carrying capacity, speed, range, cost or style, the potential for product substitution would be limited and the uptake would require some sacrifice or trade-off between different options and the environment. Of course, there are many instances in which people are willing to make such trade-offs and where all options have positive and negative features (or functionalities) needing to be resolved by the decision-maker.

3.2.2   System boundary

Whether practitioners recognise it or not, all LCA studies analyse some form of interconnected system. If there is a system being analysed, then there must be boundaries to that system. Normally, the system boundary is conceptually framed in terms of the life cycle stages included in the study. For example, a study may include all material inputs to the process but exclude capital equipment, infrastructure and services. Alternatively, the system may be described literally through a description of all the processes within the boundaries. However, as LCA can include thousands of unit processes, this is often not practical. Boundaries can be as tight as the limits of a single unit process, such as the burning of gas extracted from nature, or as broad as the consumption of goods and services by whole populations.

3.2.3   Inputs and outputs

LCAs are constructed through the calculation of inputs and outputs required, or arising as a consequence of, the delivery of the functional unit. The inputs and outputs may be technical processes such as materials, services and processes, elementary flows to and from the environment such as coal, minerals and land use, and/or inputs and outputs to air, water and soil such as carbon dioxide, nitrogen and heavy metals. All LCAs have some elementary flows, otherwise there are no impacts. The number and aggregation of technical flows vary substantially with the type of LCA and the system being investigated.

3.2.4   Impact assessment

While the type and number of indicators used in LCAs vary, all LCAs should have some indicator, otherwise analysis of impacts is not possible. Some studies claim to consider only a life cycle inventory and do not include impact assessment, although this normally means that energy and greenhouse gases are the focus of the study, or that only a very narrow group of emissions or priority pollutants are considered (e.g. nitrous oxides, sulphur oxides, hydrocarbons and carbon dioxide).

3.3   'Bottom-up' process analysis

Bottom-up process analysis refers to process-based modelling that begins at the bottom of the supply chain and pieces together the individual unit processes that make up a product's system. Traditionally, the first stage includes mineral extraction, energy production and the transport systems invariably required to produce almost anything in modern economies. For agricultural production, the first stage may be field preparation. Unique to bottom-up process analysis is that data is collected for each of these processes by measurement and modelling of each process at either local, regional or national levels, although generally the process model will represent a single process or group of processes analogous to a factory or operation. This is distinct from economic input-output analysis, where the unit processes are economic sectors. The unit processes in LCA are connected by virtue of energy and material flows between them. Hence, coal is used by electricity; electricity is used by timber milling; timber is used to make buildings, and so on. The circular nature of the economy is represented by the complicating fact that buildings are used in the extraction of coal.

One characteristic of bottom-up analysis is its focus on major materials and energy flows and the exclusion of minor and service-oriented inputs. Small material flows may be excluded, as suggested in the ISO standards, based on their mass energy or environmental significance. So, for example, where timber framing is used in coal-mining operations, it may be excluded from coal production as the impact of timber production and the mass of timber used may be less than 1% of the mass of coal extracted. In this case, timber will be insignificant in environmental terms compared to energy inputs to coal mining and transport of coal. If the LCA study is expanded so that electricity generation is considered more generally, the relevance of timber used in coal mining as an input to electricity will become even less significant. Process analysis is rich in this type of detail.

3.3.1   Process LCA: method and approach

ISO standards (ISO 14040:2006 'Environmental management – Life cycle assessment – Principles and framework'; and ISO 14044:2006 'Environmental management – Life cycle assessment – Requirements and guidelines') provide a sound overview of the LCA process, and there are many sources of advice on how to undertake LCA accordingly (e.g. 'the CML guide' to

ISO 14040 Standards (Guinee *et al.* 2001)). The following is a brief description, included here to indicate the general approach for process LCA.

The first stage is to decide on the appropriate question. Referred to as 'goal and scope' in the standards, the question asked in an LCA may be self-evident, but often requires careful consideration. The many different environmental issues and preconceptions that the people who have commissioned the LCA or the practitioners bring to the study can obscure important issues around identification of the appropriate question. Also, an initially clear question may prove to be inappropriate or inadequate, and the underlying question of the study may shift accordingly.

It is essential to plan activities and resources carefully, since data requirements often dominate the work involved in conducting an LCA. The system boundaries of the LCA need to be identified carefully, considering the implications of these boundaries for resources and data collection, the types of data to be collected, and the impacts to be assessed in the study. This may involve analysis of pre-existing LCAs of the same or similar products or different but relevant products. It may also involve conducting a small streamlined LCA (see Section 3.4) to refine the scope and identify important issues in the LCA. A process tree describing the processes inside the system boundary is also useful in the planning of the LCA.

The second stage commences with building a mathematical model of the production system. This can be as simple as a Microsoft Excel spreadsheet adding emission factors (emissions per unit of production) to the requirements for a product system. Alternatively, LCA software can be used to link thousands of processes, or a variety of hybrid approaches can be adopted. LCA modelling can be separated into the foreground system and the background system. The foreground system includes the processes that are investigated in detail. The background system includes data taken from pre-existing information, which is used in the calculation, but is not a focal point of the investigation.

A life cycle inventory for the study can then be produced. Essentially, this is a list of all resource uses and emissions that occur due to the use of materials and activities needed to deliver all options under consideration. Once this inventory is complete, the third stage involving assessment of impacts arising from these uses and emissions can be undertaken, according to the goal and scope. Assessment is typically undertaken for a set of impact indicators. These are calculated in the third stage of the LCA, usually with pre-existing factors (characterisation factors) that group resource inputs and emissions into the indicators. The number of indicators typically varies from 1 to 10 in most LCAs. Where there are more than 10 indicators, drawing conclusions can become complex, and this illustrates the nature of environmental impact assessment generally as one fraught with 'wicked' (complex) problems of relative weighting and subjective values regarding diverse and various impacts.

The fourth stage involves results interpretation. Having calculated impact indicators, LCA results should be validated in some way, and at minimum this should include a 'reality check' against expected outcomes to identify any obvious data or modelling errors. Once the general integrity of the model has been verified, results can be interrogated to identify sensitivities, major contributing factors and, in comparative studies, the key differences between the options. Once this has been achieved, alternative options may be considered, the functional unit may be varied, and extensions to the study may be undertaken.

Most LCAs use process analysis. Classic examples examined elsewhere in this book are the LCAs on waste and recycling (see Chapter 6), LCAs on water management (see Chapter 8) and the biofuels LCAs (see Chapter 10).

3.4   Streamlined LCA

Streamlined LCA encompasses a group of approaches designed to simplify and reduce the time, cost and effort involved in conducting a LCA, while still facilitating accurate and effective decisions. In a North American survey, LCA practitioners noted that a streamlined LCA:

- is simplified, pragmatic, feasible, practical, flexible, fast and easy to use
- represents the most important environmental burdens
- focuses on key impact areas
- limits consideration of effects to first-order impacts
- leaves out some life cycle stages or impact categories
- uses available information to simplify the process
- is less comprehensive
- is more 'do-able' (Weitz *et al.* 1996).

These practitioners described various approaches to streamlining LCA, usually involving one or more of the following: narrowing the study's boundaries, targeting specific issues and using readily available data, including qualitative data.

The principal question to be addressed before embarking upon any streamlined LCA involves the appropriate level of trade-off of accuracy or depth in results that is acceptable in exchange for the reduced effort in undertaking the evaluation. The colloquial term 'quick and dirty' is often applied, which sums up this trade-off. Quick and dirty LCAs invariably limit the time spent on data collection by using existing data available in public databases often already integrated into LCA software. This involves the use of data from other regions, proxy processes for data that is not available, and the exclusion of many minor processes such as material transformations, intermediate transport and so on.

Quick and dirty studies may be sufficient in themselves to address some LCA questions and can be used to scope larger projects. Skill and experience is required to understand the potential influence and significance of using data from other regions or different technological processes. Since one of the key aspects of LCA practice is learning the dynamics of the life cycle system being assessed, including how results change when parameters in the study are varied, quick and dirty LCAs can contribute to this understanding without producing an answer to the LCA question in sufficient detail or confidence for publication.

By their nature, quick and dirty studies are not usually published, but are used for internal decision-making. By way of illustration, a study was undertaken for the federal government on biodegradable plastics in 2002 (Nolan-ITU 2002). Environment Australia, in consultation with the Plastics and Chemicals Industries Association (PACIA), engaged consultants to undertake a national review of biodegradable plastics to identify and characterise emerging environmental issues associated with biodegradable plastics. Rather than conducting a full LCA, the consultants identified the various types of biodegradable plastics available, key environmental and technical issues associated with their use and disposal, particularly recycling, and a range of potential waste solutions. The streamlined LCA was then conducted on these options, modelling each of the biopolymers from available public data and studies, which in some cases were very limited. The results distinguished a significant difference between reusable 'green bags' made from polypropylene and single-use bags, but could not conclusively answer whether the biopolymer or conventional polymer bags had more significant impact, considering the limited quality of the data and the closeness and conflicting nature of the results.

Another approach used in streamlined LCA is to reduce the impact indicators and thus reduce the scope of the study and the resources required to undertake it. While reducing the indicators has the potential to reduce data collection, particularly data on elementary flows, there are two additional tasks to be undertaken. First, where indicators are to be reduced, careful consideration is required to identify those indicators of primary interest to the study so that the shortened list of indicators covers the key contestable issues. While the goal and scope can be used to exclude any indicators, the value of the study will be very limited if basic questions posed by stakeholders are not addressed. Second, at the conclusion of the streamlined study, the practitioner should comment on and contextualise the results against indicators that may be of interest but were not included in the study. This may mean referring to other studies, or considering local non-LCA data, and the relationships between indicators included and those not included, to draw inferences about expected impacts.

3.5   Input–output and hybrid input–output

Input–output analysis is a top-down economic technique, which uses monetary transactions between economic sectors rather than physical flows to represent the interrelationships between processes leading to the production of goods and services. As in process analysis, where each process can involve both direct emissions to the environment and upstream emissions in the supply processes, in input–output analysis direct emissions and resource use arising from within each sector are identified and accumulated as the necessary inputs from each sector. These are then calculated to supply final demand in any given sector. What is special about input–output analysis is that the depth of the supply chain is effectively infinite. Rather than truncating the supply chain when individual flows become seemingly insignificant, as is done in process analysis, input–output analysis effectively traces the supply chain comprehensively by resolving the infinite and circular nature of the transactions between sectors. For example, it considers the inputs from transport to make electricity, and the inputs of electricity to make trucks, and the inputs from trucks to make transport, and so on.

The limitation of input–output analysis is the coarse categorisation of economic sectors. In the 1995–96 input–output tables produced by the Australian Bureau of Statistics, the Australian economy is represented by 106 sectors (ABS 2001). The USA's equivalent input–output table includes about 500 sectors (Suh 2004), which, in terms of all the different types of goods and services produced in the world, still represents a problem of gross aggregation. Two solutions to this problem are to disaggregate the input–output data where more resolution is needed, using more detailed economic data, or to use hybrid techniques where physical flows from process analysis are integrated with the hybrid input–output data.

3.6   LCA-integrated life cycle costing

Life cycle costing (also referred to as life cycle cost analysis) is defined in the *Life-cycle Costing Manual* as 'an economic method of project evaluation in which all costs arising from owning, operating, maintaining and ultimately disposing of a project are considered to be potentially important to that decision' (Fuller and Peterson 1996). Life cycle costing is not new, although its similarities and synergies with LCA make it relevant here. Life cycle costing shares the life cycle dimension of LCA. Also, much of the technology description and flows required in LCA are also required in life cycle costing. Therefore it can make sense to undertake both evaluations at the same time. There are, however, some important differences between LCA and life cycle costing. These differences relate to the time value of money, the different perspectives of cost, and the way in which costs and prices are defined (Table 3.1).

For a detailed guide to the life cycle costing method, refer to guides such as Fuller and Peterson (1996). However, for integration of life cycle costing with LCA to provide generalised economic information for options assessed within LCA, it is worth considering the following approach.

First, the economic perspective of the life cycle costing must be established; for example, whether the costing is to be taken from the perspective of a manufacturer or other business, a consumer, or a public authority, or to encompass broader or total societal costs. Once this is determined, it will be possible to determine the point in the supply chain where costs will be

Table 3.1   Differences between life cycle assessment (LCA) and life cycle costing

	LCA	Life cycle costing
Time perspective	Long time-frame from hundreds to thousands of years considered	Impact into the future reduces due to discounting of future monetary flows
Stakeholder perspective	All environmental impacts, regardless of where they belong, are considered	Costs are associated with different players, including consumers, producers and/or societies
Data calculation	Emissions can be added up along supply chains	Cost data is often represented by price at specific points in the supply chain

defined. Different cost types to be considered include: capital expenditures, running costs, maintenance costs, replacement costs and residual value of infrastructure.

A time-frame for the economic assessment also needs to be determined. This would generally be the same as the time-frame for modelling the technical system in the LCA. Once the costs are modelled across the life cycle, they can be added after taking into account timing, expenditures and the appropriate discount rate. The discount rate can be applied when entering the cost data, taking account of it from year zero, which usually represents the current day. The alternative approach is to list cost flows with the time information so that discount rates can be applied at the impact assessment stage.

3.6.1   Life cycle costing in the water industry

Yarra Valley Water (see also Section 8.3) has experimented with the integration of LCA and life cycle costing. First, the Yarra Valley Water LCA provided typical environmental indicators including greenhouse gases, nutrient emissions and total demand on potable water supplies. Second, using the same process model, costs were calculated along the life cycle taking into account who was paying and when, so that appropriate discounting of the economic flows could be applied. The economic results were presented both as an aggregate and to show from whose perspective costs were being incurred. In the case of centralised versus decentralised water systems, it showed that costs shifted from the water authority to the end-use consumer (assuming the end-use consumer paid for on-site water systems such as greywater systems). The third step was to integrate the 'real' economic costs with the environmental indicators by monetising the environmental impacts based on current or long-term expected trading costs from greenhouse impacts, water supply and nutrient emissions. By using only a small reliable set of impacts that are currently costed and traded, the monetisation could be realised (i.e. greenhouse credits or nutrient discharge fees could be paid). The fourth step, undertaken mainly for research purposes, was to monetise health and environmental damage from air and water pollutants identified in the LCA, including air and water toxins from upstream processes that were largely caused by materials production and electricity. This gave a much higher number for the monetised value of the environmental indicators, but the costs included significant uncertainties and were difficult to realise in the short term. (For example, savings on the costs of heath damage may take many years to express themselves in a country's health system, if at all.)

So, a decision on how far to integrate LCA and life cycle costing results will depend on the time and policy perspective of the study and its stakeholders. Furthermore, like most information in LCA, the purpose of the integration should be to provide insight and clarity rather than an 'ultimate' answer to the question.

3.7   Future directions for LCA practice

There are at least two possible directions for the future development of LCA practice, reflecting the essential tension in LCA between the need for greater detail and better data, and the need for quicker, more usable tools and straightforward solutions. Accordingly, two basic directions for LCA are:

- more detailed and complicated methods
- simplified and streamlined methods.

These directions are not mutually exclusive, and indeed it is most likely that both will be pursued for different purposes. In addition, sector-specific tool integrations will become increasingly widespread in the form of assessment-tools and design-tools for specific applications. Examples where this is already happening exist in packaging design, solid waste management and building assessment (see elsewhere in this book). Typically, the tools model the technical system under consideration such as a product, building or technology configuration so that the material requirements and operational energy can be predicted and interpreted by the LCA. There is more potential to develop LCA in this way in the following sectors:

- water services planning
- product design specialties
- purchasing tools.

The scope of impacts covered under the category of 'LCA' is also continuing to expand. Integrated assessment and associated techniques will continue to extend LCA beyond the traditional functional unit approach. Currently, a pan-European project called Calcas (Coordination Action for innovation in Life Cycle Analysis for Sustainability) is looking to broaden the applicability of LCA and improve the assessment of options, particularly in relation to sustainability assessment and the science of supply and demand relationships (Calcas Project/UNEP 2007).

The integration of additional social impacts in LCA (beyond the assessment of human health which is already common) is progressing. The United Nations Environment Programme/Society of Environmental Toxicology and Chemistry (UNEP/SETAC) working group on social indicators recently proposed that social indicators should include: child labour, living wages, freedom of association, working hours, forced labour, equal opportunity, health and safety, social benefits and security (UNEP/SETAC 2007). Although the consideration of social impacts in LCA can extend its scope, some of the quantitative approaches in environmental impact assessment can also be applied to social impact assessment. Weidema describes the use of the conceptual structure and approach in life cycle impact assessment to link inventory items along a causal pathway to end-point damages (Weidema 2006). Clearly there is still some way to go in both the development of the indicators and the data, and the building of consensus around how such assessment can be used.

Open source and internet-based LCA tools will also continue to develop and expand in usage, reflecting the growth of the Internet and the multitude of needs for environmental data arising through energy, water, greenhouse and pollution reporting schemes, including voluntary reporting using LCA data by businesses. Further discussion of this and the aforementioned future directions for LCA practice are examined in more detail in Chapter 12.

3.8   References

ABS (Australian Bureau of Statistics) (2001) 1996–97 5215.0 Input-output tables, product details. Australian national accounts. Canberra.

Calcas Project/UNEP (2007) *Calcas (Coordination Action for innovation in Life Cycle Analysis for Sustainability)*. Retrieved 14 November 2007 from <http://www.calcasproject.net>.

Ekvall T and Weidema BP (2004) System boundaries and input data in consequential life cycle inventory analysis. *International Journal of Life Cycle Assessment* **9**, 161–171.

Fuller SK and Petersen SR (1996) *Life-cycle Costing Manual for Federal Energy Management Program*. Prepared for US Department of Energy, Washington DC, USA.

Guinee JB, Gorree M, Heijungs R, Huppes G, Kleijn R, de Koning A, van Oers L, Sleeswijk AW, Suh S and de Haes HAU (2001) *Life Cycle Assessment. An Operational Guide to the ISO Standards*. <http://www.leidenuniv.nl/cml/ssp/projects/lca2/part1.pdf>.

ISO (2006a) ISO 14040:2006 'Environmental management – Life cycle assessment – Principles and framework.' International Organization for Standardization, Geneva.

ISO (2006b) ISO 14044:2006 'Environmental management – Life cycle assessment – Requirements and guidelines.' International Organization for Standardization, Geneva.

Nolan-ITU Pty Ltd (2002) 'Biodegradable plastics – developments and impacts.' Report for Environment Australia, Canberra.

Suh S (2004) 'Materials and energy flows in industry and ecosystem networks. Life cycle assessment, input-output analysis, material flow analysis, and their combinations for industrial ecology.' Institute of Environmental Sciences (CML), Leiden University, Leiden.

UNEP/SETAC (United Nations Environment Programme/Society of Environmental Toxicology and Chemistry) 29 August 2007. *Second Phase of the Life Cycle Initiative. LCM Life Cycle Management Conference*. Zurich, Switzerland. Retrieved 20 November 2007 from <http://www.lcm2007.org/sessions/LC%20Initiative.pdf>.

Weidema B (2006) The integration of economic and social aspects in life cycle impact assessment. *International Journal of Life Cycle Assessment* **11**(1), 89–96.

Weitz KA, Todd JA, Curran MA and Malkin LJ (1996) Streamlining life cycle assessment. Considerations and a report on the state of practice. *International Journal of Life Cycle Assessment* **1**(2), 79–85.

Chapter 4

Life cycle assessment as decision support: a systemic critique

Tim Grant and Fran Macdonald

It is often stated that LCA is a 'decision-support tool'. Certainly it can be a powerful and systematic tool that provides useful data to facilitate decisions. But what is it about LCA that contributes to decision-making, and can it do more than support the evaluation of a previously formed proposition, as the term 'decision-support tool' would suggest?

4.1   Introduction: the role of LCA in decision support

LCA and indeed all decision-support tools play a much deeper role in framing the questions asked in order to make decisions and therefore in delimiting solutions. Fundamentally, this is due to the reflexive nature of methods of analysis on the things they analyse. The result of any analysis depends on the tool used. For example, an analysis that takes into account only quantifiable data will produce quantifiable results; and an analysis that does not consider uncertainty will not identify uncertainty in the results.

Tools that do not take into account their own reflexive nature (i.e. that they have a role in shaping the results rather than just identifying them) can be referred to as 'dualistic' – exhibiting 'the capacity to see ourselves as actors in/on the environment' and essentially separate from it (Fisher 2006, p. 4). Dualistic approaches consider how to 'fix' or cure an already defined and isolated problem. Fisher says:

> These involve isolating some 'causative' agent in the phenomenon of concern and
> 'fixing' it by altering the agent so that its outcome is not what concerned us, or by
> breaking its chain of causation so that it immobilises in the sense we understood it
> to be mobile' (Fisher 2006, p. 6).

The fix or cure is limited by the initial definition of the problem or cause. The fix may also have unintended and even undesirable consequences that are not taken into account because the initial problem has been isolated from surrounding data. Consider the infamous example of cane toads, deliberately introduced into Australia in 1935 to control infestations of beetles that were destroying sugar cane crops. The problem, defined as 'how to kill cane beetles', seemed to be readily answered by cane toads, which were native to Central and South America where their habitat was reasonably confined. However, the introduced toads had no effect on cane beetles in Australia, which persist as a problem for the sugar cane industry despite the subsequent development of pesticides. Cane toads also quickly became a pest animal, breeding unseasonably and spreading throughout the habitats of northern Australia, where they threaten wildlife that has not evolved to resist them.

A dualistic approach is distinct from a systemic approach, which recognises that the phenomenon under consideration is a part of a more general system and indeed system of systems. For example, every organism is in itself a system and an embedded part of other systems that include cells, organs, organisms, populations, communities, ecosystems, social and industrial systems and, more generally still, language and thought systems. Complex Systems Theory recognises that connections between systems are categorised by self-organisation, non-linearity, uncertainty and unpredictability (Kirshbaum 2002). Accordingly, it is reductionist and can actually lead to significant error – witness cane toads – to think of a complex phenomenon in terms of isolated cause and effect. In managing the phenomenon, it is more useful to think of its systemic context. The context is what defines the phenomenon.

A systemic approach is also self-reflexive: it recognises its own reflexivity – its own power to define the problem. This is useful precisely because one can knowingly alter the context or definition of a problem to take into account more general systems and thus gain access to a wider pool of solutions and even solutions that prevent the initial problem from arising.

This is not to say that a dualistic approach – a consideration of linear cause and effect – is 'wrong'. According to systemic thinking, a cause and effect analysis may be one useful way of construing relationships and interactions, although it cannot of itself reveal an objective truth, any more than any other method of analysis can. The types of phenomenon most suited to cause and effect analysis are 'simple' ones, where the interactions can be clearly delineated and isolated from other factors. However, environmental problems are not simple – they are complex. It is axiomatic that dealing with an environmental problem in one arena will have consequences in other arenas, and many of both the consequences and the other arenas will be unpredictable. Approaches that seek to foster environmental sustainability, like LCA, must take this complexity into account, or risk failing to meet their objective.

4.2   An analytical approach to sustainability

There are two main ways to approach the task of investigating options for optimising environmental sustainability of a product, service or system. An analytical approach examines the system, identifies environmental impacts and then tries to address them. An envisioning approach first examines what sustainability might mean, or what it might look like in the system, and then casts back to the current situation to identify ways to progress from this point to a sustainable future.

The Natural Step Framework is an envisioning approach. When The Natural Step organisation is engaged by a client, it uses a process called 'backcasting' to assess environmental impact and sustainability prospects. This involves four steps, in order:

- defining a 'framework for sustainability' of the product, service or organisation being assessed, according to a set of ecological first principles or conditions. This is where the context of the problem is identified.
- analysing the present circumstances in relation to the 'framework'
- envisioning future sustainable scenarios
- identifying strategies to fulfil the envisioned sustainable future.

The Natural Step Framework is consciously based on systemic thinking, specifically recognising the interconnectedness of environmental phenomena and their unpredictability (The Natural Step 2003).

However, 'traditional' LCA takes an analytical approach. In an LCA, a product, system or organisation under analysis is first designated as environmentally unsustainable or, at least, as having environmental impacts that may be challenged. The LCA is then used as a tool to assess and recommend ways to improve the environmental performance of the product or system. Alternatively, the LCA may be used as information by consumers seeking to evaluate alternative ways to gain the service provided by the product or system. Reports in consumer and general media of LCAs on different nappies are an example of this type of information that consumers – parents, in the case of nappies – use to make product choices (e.g. Aumônier and Collins 2005).

A great strength of LCA is that it provides definite quantifiable results, and where it reveals linear causes of impacts, it can also reveal exactly the points where change is required to 'fix' them. As such, it provides a direct conduit to decisions and taking action: it tells us 'what to do on Monday' (Kelly 2006). However, an open question remains as to whether LCA is adequate to meet the complexity of environmental issues across all cases. Does it provide solutions to systemic problems where the causes and effects are not linear but iterative, interconnected, uncertain and even, possibly, boundless?

4.3   Problem definition in LCA

LCA implicitly recognises that problems under analysis exist in systems. This is clear from the way LCA is used to define system boundaries and systems dynamics such as average or marginal supply, system displacement effects, and the interactions between the natural and technological systems involved in the analysis.

However, there is a constraint in the way LCA is often applied that potentially reduces it to a mechanistic tool, restricted to calculating the impacts of simple, isolated problems. In these cases, the initial question that the LCA is designed to evaluate is already defined and isolated by a producer, client or consumer before the LCA commences. This is an obvious corollary of the LCA proponent's interest in his or her own particular identified 'problem'. However, as stated above, a narrowly defined problem or question leads to a correspondingly narrow set of results. Hence, a narrowly set question in LCA can lead to narrow, mechanistic and sometimes inappropriate conclusions.

For example, in an LCA conducted in Melbourne in 2003 (Boyapati 2003 *et al.* cited by Public Transport Users Association), the objective was to investigate options for reducing greenhouse gas emissions of urban transport. The LCA was conducted around a relatively narrow definition of the problem: the emissions of commuter vehicles. The greenhouse gas emissions of trams were compared with those of cars and buses, and the LCA's results purported to show that a tram produces more emissions than a car, per passenger kilometre. The LCA practitioners accordingly recommended that cars were a better option than trams for reducing greenhouse gas emissions. If this recommendation had led to trams being phased out in favour of cars, it would have significantly increased greenhouse gas emissions overall. The LCA apparently did not take into account the embodied energy in Melbourne's existing extensive tram infrastructure together with rolling stock, or that this would have to be demolished and replaced with road infrastructure and new cars to accommodate a change in policy that favoured cars, or any trend to increase private car usage which may attend an increase in road infrastructure. The LCA may have provided results for the given question in simple terms, but did not recognise or consider the context of the problem – that is, the urban fabric of Melbourne.

Assessments of biofuels as an answer to energy scarcity and greenhouse gas emissions associated with fossil fuels may also founder for similar reasons. If these assessments are limited to the question of how to supply alternative energy sources to fossil fuels, their recommendations may actually create new environmental pressures. A significant increase in biofuels production may lead, for example, to such competition for land use that food crops are held back in favour of fuel crops, resulting in increased food scarcity (Beer *et al.* 2002). The extent to which such outcomes can be foreseen using LCA is, of course, a function of the appropriate scoping of the study, data and modelling, and constraints of the technique. For example, the study by Beer *et al.* (2002) did not take into account land use or water use, as is the case with most LCA work on biofuels to date internationally. In this case, the practitioners recognised the omission as a constraint on impact identification.

At first glance, it appears that the pitfall of the already defined question can only be avoided by explicitly defining or re-defining the question in its systemic context as a first stage within the LCA. There is an obvious and fair objection to this. Naturally, clients engage LCA with questions they have already identified, which they need assistance to resolve. If these questions were not acceptable, LCA would have few clients. That there are boundaries to the questions is also practical, given limited time and resources, let alone that a bounded question is necessary to give the process any direction.

The pitfall is not, however, automatic. It can be resolved to the extent to which the LCA is self-reflexive; that is, to the extent that it recognises that the results are determined by the scope, method and limitations of the assessment itself. LCA can achieve this in the fourth stage, 'Interpretation', where the results of the inventory analysis, duly related to indicators in the impact assessment, are examined to assess whether they provide a sufficient level of evidence to draw valid conclusions. The results are tested against each of the three previous steps using a variety of methods, including sensitivity analysis, completeness checks and data quality analyses. In referring the results back to the first stage, 'goal and scope definition' (see Chapter 3), the objective of the initial question is revealed and any potential conclusions from the results are examined to see whether they meet that objective. It may be shown, at this stage, that the objective is not met by the results because the initial question has not been designed to achieve this.

A well-known example where this process resulted in a recommendation to re-define the original question was the second major LCA ever conducted in Europe, on biopolymers in the 1970s (Oberbacher *et al.* 1996). The initial question was about the difference in environmental impact between plastics made from biopolymers and other plastics, the objective being to recommend ways to reduce the impact of plastics. The results showed that both sets of plastics involved considerable environmental impact, and therefore answering the initial question would not lead to any significant reduction in impact. So the recommendation was that the more general role of packaging and plastics in the economy should be challenged to see how their impact could be tackled. Based on a consideration of the context of the problem – the flow of plastics through the economy – the final recommendation addressed deeper structural causes of environmental impact and suggested more far-reaching solutions than the initial question would have allowed.

The following two Australian LCA examples also illustrate how interpreting the context of the problem or initial question can lead to systemic solutions.

4.3.1   Case study: Yarra Valley Water

Yarra Valley Water, a Victorian water authority, commissioned an LCA to compare the authority's reticulated water supply system with small household rainwater tanks, to see whether the tanks should be introduced to save water (Grant and Hallman 2003). The results showed that the tanks did not compare well with the existing reticulated system. On interpreting the results, it was realised, however, that it did not make sense to recommend extending the reticulated system as an alternative to tanks. The existing infrastructure was fixed and could not be extended, and the results did not point to the existing system being the best available way to save water. Instead, the initial question was re-defined to take into account the authority's objective, and a further LCA compared a range of water-saving mechanisms as possible alternatives to reticulated water.

4.3.2   Case study: Yalumba

Yalumba, a wine producer based in South Australia, commissioned a quick assessment to compare the environmental impact of different types of wine casks. However, the results led to the question: what is the environmental impact of wine bottles compared to casks? Accordingly, the LCA was extended to compare bottles and casks. Although the results showed that the impact of bottles was greater than casks per litre of wine delivered to the customer, a result magnified by bottles having less volume than casks, it could not automatically be concluded that casks were therefore better than bottles because this did not take into account the context of the market for Yalumba's wine. In this context, bottled wine is a premium product and not generally drunk in the same way or for the same reasons as cask wine. Consumers choose between bottled and cask wine on the basis of their budgets and the particular 'wine experience' they seek, and the quantity of wine delivered by each packaging type is generally less important in this choice. The 'wine experience' includes a whole raft of non-material factors, such as the sense of occasion provided by a bottle as opposed to a cask, perception of wine quality, and so on. One of the most important factors is that bottled wine may be bought as part of a collection to be stored ('put down'), whereas cask wine is never bought for this reason as its shelf life is only about six months. Because bottled wine is usually more expensive than cask wine for all the reasons given above, consumers also buy less wine when they buy bottled wine. The functional unit of the LCA was re-defined to reflect this context and became A\$50 worth of wine at retail (gross turnover) rather than one litre of wine. On this basis, the LCA results showed that the impact of bottles was actually similar to that of casks. When the impact of the wine itself was also taken into account, the results showed that the impact of wine in bottles (wine plus bottle) was less than half that of wine in casks.

4.4   Focus on functionality

The Yalumba example illustrates another way in which LCA can usefully engage with the complexity of environmental problems. This is LCA's focus on functionality. One fundamental context of any product, system or organisation under assessment is its function or utility. This function, however, is multi-faceted and may be variously defined. The functional unit in an LCA is set according to the defined common function of the things being compared. How that function is defined greatly influences the results.

An examination of functionality in the following example makes this clear. In Europe in the 1990s, several environmental and economic studies were undertaken to evaluate alternative ways to minimise the energy consumption of clothes washing machines. The studies concluded that there was little scope to improve energy efficiency because technical efficiency of washing machines was already high, although there was some scope to decrease energy consumption by addressing consumer behaviour, namely textile care in the home, especially detergent use, load levels in the machine, drying and ironing practices. The functional unit for these studies could be said to be 'cleaning clothes with a home-owned washing machine'. A pilot scheme undertaken by Electrolux in Sweden focused on an expanded function of 'cleaning clothes in the home' (Jones and Harrison 2002). In this scheme, Electrolux and a local energy utility offered customers a pay-per-wash option. Customers had a washing machine at home, but did not own it and instead paid for their number of usages. This gave them an incentive to reduce the number of usages and so save energy. If the function is further generalised as, for example, 'cleaning clothes', non-domestic alternatives such as commercial laundering could be considered. If the function is even further generalised as 'providing adequate clothing', then alternative fabrics and alternative garment designs, along with issues such as wear resistance, moisture and odour absorption or control, would all be relevant. Consider, for instance, 'thermal comfort' as a function of clothes. In colder climates, the health benefits of warm clothes could be taken into account to evaluate the reduction in human health damage provided by giving warm clothes to poor and homeless people. In warmer climates, the role of fabric in blocking exposure to UV light could be taken into account. In this case, clothes could be compared to umbrellas or sun screens.

Getting to the heart of the reasons people purchase and use different products and services can open up new opportunities for resolving the environmental problems associated with them. It is also clear that context reveals possibilities for impact outcomes that would otherwise be unexpected, and may lead to new options being contemplated. Whether or not these new options are practical as well as possible, they tend to transcend those considered within the original, narrower context. For example, the question of the technical efficiency of the spin cycle on washing machines is of little relevance for clothes made from materials that do not require to be washed with a vigorous spin cycle.

4.5   Challenging consumption

The downside of the focus on functionality is its potential to justify consumption. By understanding *why* we use products and services, *whether* we need to use them in the first place can be obscured. This is related to LCA usually examining the relative performance of products or services within a consumption bracket, and so rarely considers the overall scale of environmental impact of those products and services. The effect of this focus can been seen in suppliers' and consumers' responses to eco-labels and other 'green' labelling schemes, which are usually based on LCAs. An eco-label sends a clear message – it is designed to send the message – that purchasing the product so endorsed will be relatively good for the environment. Such a message conspicuously fails to address the possibility (or even probability) that not buying any product in that category may be even better for the environment. For example, a clothes dryer with a relatively high star rating may compare favourably with other dryers in a whitegoods store, but that same store will not sell, let alone promote, washing lines as an alternative to its range of dryers.

Another constraint on the use of functionality is that, while common function is explicitly defined as part of an LCA, once the functional unit itself is established, it is the one constant against which everything is measured – and is thus rarely challenged. There have been some attempts in LCA to address macro effects or scale of consumption, in particular increases in consumption due to increases in efficiency and corresponding decreases in cost. This up-scale of consumption, known as the rebound effect, offsets projected reductions in demand for energy and material resources that occur as a result of improvements in technology efficiency and green product development, but also as a result of LCAs and eco-labelling. Direct effects occur when consumers choose to use more of the efficient product because it is cheaper. For example, a more fuel-efficient car may be driven further or more often if the cost of fuel is normally a limiting factor in car use. Furthermore, indirect effects occur when consumers use the money they have saved on the more efficient product to buy more products. Market-wide effects occur when the decreased cost of using a resource, for example electricity, opens up economic opportunities for its use in new products. With regard to electricity, rebound effects have been shown to reduce projected electricity savings by 10% to 40% depending on the product (Gottron 2001; Herring 2006).

One response to such macro effects of consumption is the 'E2 vector' (Goedkoop *et al.* 1999). This is a European development in LCA that allows identification of both the absolute environmental impact and the absolute economic cost of alternative options. The objective here is to reveal the effects of economic growth and environmental load, and the links between them, including the rebound effect. Reductions in environmental impact can then be contextualised in terms of decreases in cost to the consumer. The basic premise is that until it can be established how money savings are spent, the net effect on the environment of substituting a more efficient product for a less efficient one cannot be determined.

The CSIRO has developed the Australian Stocks and Flows Framework, a computer model of the Australian economy based on physical stocks (e.g. mineral resources, water, soils and forests) and flows (e.g. food and waste) and their interaction with demographic and economic factors. By tracking total material stocks and flows of materials, goods and people in Australia, this model can be used to determine the limits to scaling-up a material – or the limits of product substitution – and also the indirect physical effects of substitutions. Present situations can be modelled, but the future can also be projected based on expected population growth and demographic and geographic shifts. The limitation of this tool is that it models at a relatively coarse level: while it considers macro-level factors, it does not consider specific products or technologies. For instance, it includes an 'appliances' category, but not a comparison of different types of appliances within that category, such as different types of washing machines.

4.6   The non-result in LCA

The second way in which LCA helps to challenge consumption is by being clear about what it does not do. LCA can be used to examine all alternative approaches to deliver given functions, which may show that all options have similar impacts – or that no option offers the 50% to 90% reduction in impact which may be expected from a sustainable alternative. In particular, in consequential LCA, where the effects of changes in supply and demand are modelled when introducing a new function or service into the economy, LCA can identify that consumption of even low-grade waste products may have flow-on effects and impacts elsewhere in the economy. This result in LCA is often seen as a non-result with no option that makes a discernable difference on environmental impacts compared to existing practice. This is essentially a proof that the level of consumption is unsustainable and that product substitution and cleaner, more efficient technology, of themselves, will not make any significant difference. The European biopolymers study and the first Yalumba cask study show how useful this non-result can be.

4.7   Revealing the world behind the product

The United Nations Environment Programme (UNEP) has referred to LCA as a tool for revealing 'the world behind the product' (Fava 2002). This encapsulates arguably one of the most significant potential contributions LCA can make to our understanding and management of sustainability. Beyond simply expanding the knowledge base, LCA can also reveal blind spots in the knowledge base, and areas of ignorance about environmental impact. Damage incurred wilfully or through negligence is at least subject to checks imposed by law and public pressure. However, damage incurred through ignorance continues unabated until the ignorance is exposed or the system collapses. Accordingly, revealing areas of ignorance opens up opportunities for great advances in sustainability.

LCA can illuminate the 'depth' of supply chains, and their integrated and connected nature. It can also help break down arbitrary distinctions between natural and unnatural systems because it calculates environmental impacts on the merits of standard impact metrics. Hence, while agriculture is a 'natural' system, LCA studies generally indicate that agricultural systems have higher impacts than other obviously 'unnatural' systems, such as refineries. Of course, the scope to reveal these understandings is limited, particularly for consumers, where what is emphasised or publicly revealed in the results of an LCA is merely a final score. In participative

LCA, where the process of the assessment is revealed and emphasised, the educative value of the LCA may be far greater (e.g. see Norris 2005).

4.8 Conclusions

The successful application of LCA depends on the appropriate question being asked, and a wide understanding of the context and meaning of the outcome. Although it is often applied to addressing 'big' questions, it is important that the results are not seen as fundamental 'answers'. First, an LCA is a snapshot of a product system at a point in time under specified assumptions. It generally has little to say about the adaptability of the system, its limits, risks or potential. Also, new innovative technologies often look inefficient in the early design stage and can fare poorly in LCA terms even if they are potentially of great benefit to the environment. More importantly, the results should never be taken literally or in isolation from the broader dynamics and potentials of the product system and human system in which the things being assessed occur. The knowledge 'frames' or paradigms of the different stakeholders involved are much more influential in sustainability decisions than any supply chain identified by LCA (Tukker 2000).

Products are embedded in strong cultural, economic, legal and political structures and systems, which variously facilitate or impede proposed environmental solutions (Fisher 2006). The practicality of proposed solutions depends on these structures, systems and the interrelations between them. Consider, for example, the political objections to challenging the role of plastics in the economy, as opposed simply to proposing biopolymers. Various organisations exist purely to promote plastic products. At the Australian LCA Society (ALCAS) conference in 2006, the American Plastics Council, which actively uses and promotes LCA, stated that its mission is to 'make plastics a preferred material, helping to defend and expand market opportunities for its businesses' (Lew 2006). Approaches that take account of existing 'frames' and structures, particularly those approaches that allow participation of stakeholders and negotiation, provide more potential for a new consensus and a shared vision – and even new transformative pathways for change.

Finally, there is a tension between LCA not in itself providing comprehensive, objective or 'big' answers, yet having a role in producing unique, profound, useful information, which can lead to fundamental shifts in practice. While LCA results do not themselves provide solutions to sustainability questions, they do increase the knowledge base, which can reveal pathways to sustainable solutions. Furthermore, an LCA conducted with integrity and based on systematic calculations produces results that are scientifically robust and reliable.

So, what does this tension say about how the development of LCA may proceed? One option is to constrain its application. LCA practitioners could 'acknowledge that they can't come up with overall answers and just produce a tool that generates limited data' (Tukker 2000). This approach suggests a new repression of LCA. There is, however, an alternative that is 'braver' (Goedkoop and Alvarado 2006) and engages both with the urgency of sustainability and the great potential of LCA to help achieve it. When Dutch designers and LCA practitioners came together in the early 1990s to develop an eco-indicator, the designers insisted that a single indicator was necessary while the LCA practitioners insisted that it was impossible, given the complexity of the data. After six months of meetings, they realised they were both right, so they went ahead and developed an indicator anyway (M. Goedkoop, pers. comm.). This approach acknowledges that LCA cannot deliver objective answers, let alone the 'right' answers, but it can generate effective answers, and we have a responsibility to come up with the best answers possible considering that the world's environmental problems require them. Taking this view, it would be a mistake to constrict the use of LCA. Instead, the methods and tools should continue to be explored and evolved.

4.9   References

Aumônier S and Collins M (2005) *Life Cycle Assessment of Disposable and Reusable Nappies in the UK*. Environment Agency, Bristol, UK.

Beer T, Grant T, Morgan G, Lapszewicz J, Anyon P, Edwards J, Nelson P, Watson H and Williams D (2002) 'Comparison of transport fuels: life-cycle emissions analysis of alternative fuels for heavy vehicles.' CSIRO, Aspendale, Australia.

Boyapati E, Hartono A and Rowbottom J (2003) Comparison of emissions from public transport and private cars. In: *Proceedings of the 8th Cairo International Conference on Energy and Environment*. Egypt. In: Public Transport Users Association, Common urban myths about transport. <http://www.ptua.org.au>.

Fava JA (2002) Life cycle initiative: a joint UNEP/SETAC partnership to advance the life-cycle economy. *International Journal of Life Cycle Assessment* 7(4), 196–198.

Fisher F (2006) *Response Ability: Environment, Health and Everyday Transcendence*. Vista Publications, Melbourne.

Goedkoop M and Alvarado C (2006) Three solutions for weighting across impact categories. In: *Proceedings of the 5th Australian Life Cycle Assessment Conference*. Melbourne. Australian Life Cycle Assessment Society (ALCAS), Melbourne.

Goedkoop MJ, v. Halen CJG, te Riele H and Rommens P (1999). Product service systems, ecological and economic basics. PRe Consultants, Amersfoort, The Netherlands.

Gottron F (2001) 'Energy efficiency and the rebound effect.' CRS Report to Congress RS20981. National Council for Science and the Environment, Washington DC.

Grant T and Hallman M (2003) Urban domestic water tanks: life cycle assessment. *Water* **August**, 22–27.

Herring H (2006) Energy efficiency – a critical view. *Energy* **31**, 10–20.

Jones E and Harrison D (2002) Investigating the use of TRIZ in eco-innovation. The TRIZ Journal. Retrieved October 2007 from <http://www.triz-journal.com/archives/2000/09/b/index.htm>.

Kelly T (2006) The road map to a better future. In: *Proceedings of the 5th Australian Life Cycle Assessment Conference*. Melbourne. Australian Life Cycle Assessment Society (ALCAS), Melbourne.

Kirshbaum D (2002) Introduction to complex systems. *The Complexity and Artificial Life Research Concept for Self-Organizing Systems*. <http://www.calresco.org/intro.htm>.

Lew M (2006) Plastics and the US LCI database project – strategic drivers. In: *Proceedings of the 5th Australian Life Cycle Assessment Conference*. Melbourne. Australian Life Cycle Assessment Society (ALCAS), Melbourne.

The Natural Step (2003) *The Natural Step Framework*. The Natural Step Environmental Institute Australia, Melbourne. <http://64.207.158.76/au.naturalstep.org/contacts/contform.html>.

Norris G (2005) The contribution of research in dialogue with reflective practice for sustainable consumption. Contributing Paper, *Proceedings of the 4th Australian Life Cycle Assessment Conference*. Sydney. Australian Life Cycle Assessment Society (ALCAS), Melbourne.

Oberbacher B, Nikodem H and Klopffer W (1996) LCA – how it came about. An early systems analysis of packaging for liquids. *International Journal of Life Cycle Assessment* **1**(2), 62–65.

Tukker A (2000) Philosophy of science, policy sciences and the basis of decision support with LCA. *International Journal of Life Cycle Assessment* **5**(3), 177–186.

Chapter 5

The Australian environment: impact assessment in a sunburnt country

Tim Grant

5.1 Introduction

There is a substantial body of ecological and climatological evidence that the landscape and environments of Australia differ significantly to those of other continents. Australian ecosystems are in many ways distinct, due to the size, physical isolation and natural climate variability of the continent. Varying cycles of rainfall along with land use practices lead to episodic and endemic problems with salinity and eutrophication. Meanwhile, about 80% of Australian plant species and vertebrate animals are unique in the world (SoE 2006).

This presents challenges for life cycle assessment (LCA) because, in order for LCA findings to be used confidently and applied beyond the specific case, the technique relies on the subject(s) of enquiry exhibiting a range of properties, including reliable 'representation of the average' around a normal distribution. LCA also relies on the availability of inventory data and appropriate impact assessment algorithms. In a widely varying environment, which is different from the global 'norm' (if this exists), and in a situation where most of the inventory data and impact assessment algorithms have been developed in Europe or North America, there is an apparent difficulty for LCA practice in Australia.

The issues associated with LCA methods, applicability, transferability of results, inventory and impact assessment have been introduced in Chapters 2–4. The focus of this chapter is the 'uniqueness' of the Australian environment, the challenges that this presents for LCA practice, and the ways in which these challenges can, are and may be met. In Section 5.2 some of the 'unique' characteristics of the Australian natural environment are introduced, and in Section 5.3, features of the urban environment. In Section 5.4 the concept of bio-productive capacity is introduced and considered in the Australian context. In accepting the 'unique' characteristics of Australia, what these differences mean for LCA is probed in Section 5.5 – in both undertaking LCA and in interpreting the results.

5.2 Land use and the natural environment in Australia

The continent's terrestrial ecosystems fall broadly into the following categories: rainforests, sclerophyll forests, savannah woodlands, grasslands, deserts and freshwater wetlands, but the overarching physical conditions are dryness, and variable rainfall and river flow (Archer and Beale 2004). Australia is the driest inhabited continent on Earth. The State of the Environment Report 2006 details the current physical state of Australia's settlements and landscapes (SoE 2006) and a short summary of some key aspects is pertinent here.

Most people live close to the coast, with a trend away from rural communities towards interlinked coastal cities. Urbanisation is an ongoing phenomenon, and there is a link here to losses of the natural environment, since most population expansion occurs in greenfield developments in the outer suburbs of larger cities (see Section 5.3). Hence, coastal environments are subject to significant pressures, and the rate of natural resource loss in these (often sensitive) transition ecological zones is therefore a significant and ever-pressing issue.

Across the country, agriculture accounts for 62% of land use. Sheep and cattle grazing contributes in particular to loss of habitat and soils in more fragile arid ecosystems. Soil acidity due to planting of legume-based crops, and soil salinity due to irrigation, are also significant problems. Most of Australia's original vegetation remains, but in many areas is in decline, due to clearing for agriculture, changed fire regimes, introduction of invasive species, and disruption of understorey in forests and woodlands. Australia's remaining forest areas cover only about 15% of the land they covered when European settlers arrived. Clearly there is tension over land use in Australia, with demand for agriculture, wood production, natural catchments, mining and urban development.

5.3 Human settlements and demand for energy and water

In contrast to the natural environment, settlements in Australia are comparable to those in other modern westernised countries. Most Australians lead highly urbanised lives situated in large cities that broadly resemble cities in Asia, Europe and the Americas, albeit with distinctive aspects to the urban form, dynamics of development and demographics. Melbourne has a 'European' core, with walkable, mixed use, and vertical as well as horizontal streetscapes, offering '24/7' services combining indoor/outdoor dining, and recreation, as well as retail, office and commercial spaces. Attempts to integrate and encourage public transport add to the European feel, with trams and cycle lanes claiming a share of the spaces between buildings. However, the outer suburban form is more 'North American', and is dominated by low density residential 'quarter acre blocks', representing the enduring Australian dream of home ownership in a sprawling, home/backyard, private car-based environment. While there is space here for trees and ribbons or vestiges of natural environment, this space has often been utilised as concreted access or yard areas, lawns on the 'English' garden model, and exotic garden planting, requiring considerable watering during dry periods.

The areas of Australia that have seen greatest development in recent decades are on the eastern seaboard, while in the west, suburbs north and south of Perth are also expanding rapidly. Australia-wide, some 77% of the coastal development is pre-1980 (SoE 2006); this figure also indicates that almost one-quarter of coastal development has taken place since 1980 – and the 'sea-change' pressure continues, as baby boomers and their offspring seek out the seaside. Indeed, it is predicted that some 9.2% of the total Australian coastline is likely to be developed by 2050 (SoE 2006).

Meanwhile, Australia's population has increased by about one million since 2000, with an annual growth rate of 1.2%. Australia's growing cities are taking over productive agricultural lands and areas of heritage and ecological significance, and this pressure is accentuated by increasing consumption of energy, land, water and other products dependent on natural resources. Demand for water, exacerbated by the recent drought, is placing significant pressure on Australia's inland water systems, particularly through the use of groundwater. A national water policy reform process, initiated in 1994, is being implemented. This will eventually significantly change the way water is managed in Australia and is likely to have a major impact on water consumption and the health of waterways. Meanwhile, water consumption continues to increase. Irrigated agriculture accounts for about two-thirds of Australia's water use, while direct domestic use accounts for only about 9%. However, these agricultural practices are driven by the dual needs of food consumption (i.e. human settlements) and export-led economic performance. The use of rainwater tanks is increasing, although there is still little reuse or recycling of sewage effluent or stormwater. Nearly 50% of household water is used in gardens.

Energy use is increasing and is reliant on fossil fuels such as coal and oil, with little take-up of renewable energy. Electricity consumption in particular is increasing, partly due to a growth in commercial and residential air conditioning of 20% per year since 2001. Road transport consumes almost 40% of all energy used in Australia, and private passenger vehicle travel represents three-quarters of total road travel. Despite recent attention to urban design to mitigate energy consumption, Australian governments continue to budget for freeway construction, particularly in Sydney and Melbourne, at the expense of improving public transport. Meanwhile, in the Australian household, an enduring result of modern consumption patterns is the production of waste. Australians dispose of about 1 tonne of waste per person per year to landfill (see Chapter 6 for details on waste management in Australia).

5.4   Bioproduction and capacity in Australia

Given the uniqueness of the Australian environment, the question arises as to exactly what is the bioproductive capacity of this fragile, arid land. The limited information that does exist regarding the state of Australia's biodiversity is weighted towards coastal biodiversity due to the rising concern over recent decades about the rate of coastal development. Many threatened species are in the Murray-Darling Basin, south-west Western Australia, populated coastal regions, and in the Tasmanian Midlands, with more than half of the ecosystems in the developed coastal areas and the Murray-Darling Basin under severe pressure and significant decline (NLWRA 2002; Olsen *et al.* 2006; Tyler 2006, cited in SoE 2006). Of 901 nationally important wetlands in Australia (as at 2006), a 2001 assessment found that 231 were under pressure from changes to water regimes.

Annual average bird numbers on floodplains have fallen from 1.1 million in 1983 to 0.2 million in 2004. Four species of frog are already extinct, 15 are endangered and another 12 are listed as vulnerable, with some 14% of frog species threatened (SoE 2006).

Given this dynamic flux in biodiversity, and signs of vulnerability, the question arises as to whether human settlement is already exceeding its ecological carrying capacity. If it is not, then what additional capacity remains to enable Australia to absorb a growing population? Mathis Wakernagel's thesis on the 'ecological footprint' ('eco-footprint') (Wackernagel and Rees 1995) set out to determine the rate at which human beings are consuming the resources of the Earth and, in particular, the extent to which this rate of consumption exceeds the rate of accumulation and/or regeneration of resources. A full discussion of the eco-footprint method of calculation is outside the scope of this book. For the purposes of this discussion, however, it involves two essential elements:

- a measure of land used per person, according to the different sectors of the economy and the population's draw on these economic services
- calculation of the energy used per person, based on a similar calculation converted into an equivalent land-area unit through an algorithm based on estimates of land required to produce this quantity of energy.

Following Wackernagel and Rees' seminal work, subsequent calculations published in the Living Planet Report (World Wildlife Fund (WWF) 2006) indicate that humans are over-consuming the bioproductive capacity of the Earth by about 22%. However, beyond this 'average', there appear to be large variations between communities. Of particular relevance here are the implications for Australia. According to the WWF eco-footprint work, Australians are consuming the world's biocapacity at three times the rate at which it is being produced or regenerated. Nevertheless, the biocapacity available in Australia still exceeds our consumption: WWF states that 12.4 global average hectares of biocapacity are available per person, while only 6.6 global average equivalent hectares are being consumed per person.

The basis of the eco-footprint is aggregated and generalised data which, given the previous discussion regarding the uniqueness of many of Australia's ecosystems, may not be appropriate to be applied to those ecosystems. Notwithstanding the doubts involved in applying eco-footprint calculations, there is a strong suggestion from this work that Australia has substantial productive capacity coupled with one of the highest per capita consumption rates (sixth highest consumption footprint in the world). Add to this the drivers for competition in the global economy, and it becomes clear that Australia sits on the horns of a dilemma. The temptation to eat further into biocapacity reserves in order to boost economic performance in the short term sits uncomfortably alongside the emerging global view of Australia as an environmental pariah – over-consuming, over-polluting and generally being profligate with the resources provided by the global environment.

The position of Australia as relatively underpopulated and with available bioproductive capacity places special responsibility on policy and decision-making in Australia as to the appropriate uses of this biocapacity. The Australian economy relies heavily on the export of primary resources such as minerals and agricultural products. Any shift toward using agricultural capacity to supplement depleting fuel reserves through biomass energy technologies will affect both the domestic balance of trade and global food supply. The tension between food and fuel is increasingly recognised as an issue (UN News Centre 2007). (See Chapters 8 and 10 for further discussion.) Likewise, the balance between use of bioproductive capacity for short-term economic gain and its retention for future biodiversity needs is inevitably contested and contentious.

5.5   Discussion: implications for LCA and a sustainable Australia

Discussion so far in this chapter has focused on the distinctiveness of the Australian environment in relation to its ecology, land use and biodiversity, and the urban environment with respect to water and energy use. These can be regarded as input factors in LCA. The next issue is the extent to which impact burdens vary in Australia, compared to other countries. Then, across the input (resource) and output (burden) categories, we can contemplate the implications of this Australian 'distinctiveness' for LCA and sustainability assessment.

LCA requires that indicators of environmental impact are defined in advance of impact assessment (see Chapter 3). This means that, while it is impossible to derive an overall vision of sustainability from LCA (as from any assessment procedure), it is possible to set out a range of impact factors, which taken together may represent a suitable surrogate measure of sustainability. Indeed, one explicit aim of LCA is to avoid burden shifting between different environmental impacts or between different timeframes by choosing a range of impact factors. Hence, LCA supports the multi-issue and intergenerational equity components of sustainability, and aims to avoid narrow definitions that could lead to burden shifting. (See Chapter 4 for the burden shifting problems of narrow definitions.)

Indicator models used in LCA implicitly take a position about what should be included in environmental impacts. The Eco-indicator model developed in Europe, using a 'top down' approach (see Chapter 3 for a definiton of 'top down' approaches), began by defining what is meant by the term 'environment' (the 'Eco' we indicate). The chain begins with Areas of Protection (Udo de Haes 1999), which are ultimately conditions or values that humans wish to maintain or protect. These can be human health and/or the health of the Earth. Damage

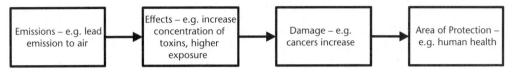

Figure 5.1   The Eco-indicator approach (after Udo de Haes *et al.* 1999).

Categories, threats to the Areas of Protection, are then defined (e.g. cancer). Damage Categories are developed from analysis of specific effects that threaten or enhance the Area of Protection, such as radiation. These effects are in turn linked to emissions, waste generation or resource use in the life cycle inventory.

Accepting the Eco-indicator approach, 'sustainability' in LCA terms means maintenance, enhancement or damage minimisation in key areas of protection. The areas identified in the Eco-indicator best practice working group include life support and resources, ecosystem health and human health (Udo de Haes *et al.* 1999). Having already introduced resources and the distinctiveness of Australian ecosystems, we can now consider the implications of these for impact assessment, both for land use/ecology and human health.

One important factor in Australia in relation to land use is water. Given the water-scarce nature of much of the Australian environment, any LCA with the aim to establish the impact of a product or service on Australian ecosystems must properly account for the inevitable competition for water within catchments. Catchments typically provide water for population settlements, agriculture and environmental flows. Approaches to land use assessment in LCA also vary from an index of use or conversion of land to an 'un-natural' use (for example, in the damage footprint developed by Lenzen and Murray (2001)) to estimates of species loss from each incremental change in land use (such as in the Eco-indicator 99 (Goedkoop and Spriensma 1999)). There is also hope that some of the biodiversity metrics, such as habitat hectares, could produce measures that are generalised enough for LCA in Australia, although this area requires development.

Many land use changes are irreversible (e.g. the destruction of native forests), and much damage to land and water systems is derived from historical actions, both of which further complicate land use impact assessment. Clearing of land, introduction of weed species and feral animals, and the modification of water tables have led to some of the largest impacts on ecosystem quality in Australia. The impacts of current production and consumption appear to contribute little in many of these historical issues or, if so, only at the margins.

Biodiversity and ecosystem diversity are site-specific and difficult to generalise into LCA. Many subtle management effects are important to ecosystem diversity and maintenance. These can include diverse factors such as fire (or the lack of it), flood regimes and protection from pests. While pressure from production and consumption of goods may affect some of these issues (e.g. timber production will preclude fire regimes), it is almost impossible to determine clear, generalisable links.

Turning to human health, work on the impacts of toxic substances in Australia has been undertaken by the University of New South Wales and Mark Huijbregts at Nijmegan University (the Netherlands). This work has examined fate and exposure factors (how emissions of toxins find their way into humans) and compared these to models developed for the USA and Europe. The dryness of the Australian continent appears to reduce people's exposure to toxic pollutants, as water is an important transfer mechanism in their movement. The fate factors for human exposure of substances emitted to agricultural soils are, on average, 160 times lower than those in Western Europe, and substances emitted to air, fresh water and seawater are 20 times lower than the average (Huijbregts *et al.* 2003). This work has also found large uncertainties in the modelling of toxic emissions in Australia.

The Australian Year Book (ABS 2004) notes:

> particle pollution is a major health concern as it can exacerbate respiratory and cardiovascular illnesses, including bronchitis, pneumonia and asthma, leading to increased hospital admissions (Atech 2001). Particles have also been linked to the deaths of up to 2400 people a year in Australia, carrying an associated cost of $17.2b (Environment Australia 2001).

Indoor air quality studies show that our building environments are very similar to those in other countries. Indoor air pollution is of particular concern as Australians spend up to 90% of their time inside (ABS 1997). While specific data on health consequences from indoor air pollution is not available in Australia, evidence from other countries shows this to be a significant impact and one that LCA should take into account.

LCA work by Grant and Beer (2000) on alternative transport fuels initially found significant impacts from particles, although many of these were from on-farm impacts and are therefore not clearly relevant as contributors to urban air pollution. In subsequent analysis (Beer *et al.* 2001), a distinction was made between non-urban pollutants, others containing those known to be in urban areas, and those for which no location information was known. A similar approach was taken by Ross and Evans (2003). The problem of tracking site-specific data is being resolved with the recent inclusion of subcompartment specification in the international standards report on data documentation (ISO 2001). This allows for air emissions to be further broken down into urban, non-urban, indoor, stratospheric, tropospheric categories, and so on. While the international standard does not contain a definitive list of subcompartments, the practice in the ecoinvent database (Frischknecht and Jungbluth 2004), and implementation of this in LCA software, has settled on five subcompartments in both water and air, which include urban and non-urban separation. Although this information is available, it is only of use if background databases utilise this feature and collection of data includes this subcompartment specification where it is required.

5.6   Conclusions

Two key issues arise out of the uniqueness of Australia as a place to conduct LCA. First, the use of international data is fraught and should be undertaken with special caution. Second, there are unanswered questions about the implications for any weighting of impacts.

Regarding the first issue, because the Australian environment is unique and deserves special attention in LCA, it raises problems for impact assessment for particular impact types. Australia's distinctive ecosystems and biodiversity determinants require special care in translating impact assessment models developed in Europe or North America. (See Chapter 8 for further discussion of Eco-indicator sets for Australia.) On a positive note, since Australian people are similar to other populations, the model used for human health impacts in Australia can use existing models modified for local climatic and population dynamics. Similarly, global impacts are characterised by the source of the emissions being irrelevant, at least in terms of geography. These impacts include global warming, ozone depletion and climate change. There is no specific reason to modify these impact models to Australia, as the current models incorporate all global impacts including those in Australia.

Regarding the second issue, notwithstanding that some impact factors can be translated into other factors in LCA, the method requires proper assessment of a large enough range of impacts if the result is to reflect a sustainability measure. This will reduce the potential for burden shifting. It occurs, for example, where global climate change impacts are counted, but not biodiversity impacts; therefore, decisions are made that are detrimental to biodiversity on the pretext of reducing climate change impacts. Because LCA has been developed and adapted from Europe, the impact models have not included proper evaluation of land use in Australia. Water extraction is also poorly characterised. Unless impact categories are specifically developed to account for these factors in Australia, unsustainable burden shifting will occur in LCAs undertaken in this country. Certainly, before 'total environment impacts' can be characterised for human activities in Australia, significant research is required to develop a logical set of impact factors that explain the relative importance of water, land use and other unique impacts in Australia within the full range of impacts. The challenge, as discussed further over subsequent chapters, is therefore to develop LCA data, algorithms and methods that adequately reflect the Australian environment.

5.7   References

ABS (1997) *How Australians Use Their Time*. Australian Bureau of Statistics, Canberra.

ABS (2004) *Australian Bureau of Statistics Year Book Australia 2004*. ABS Catalogue No. 1301.0. Commonweath of Australia, Canberra.

Archer M and Beale B (2004) *Going Native*. Hachette Livre, Sydney.

Atech (2001) *Woodheater Emissions Management Program for the Tamar Valley-Scoping Study*. Environment Australia, Canberra.

Beer T, Grant T, Morgan G, Lapszewicz J, Anyon P, Edwards J, Nelson P, Watson H and Williams D (2001) 'Comparison of transport fuels – final report to the Australian Greenhouse Office on the Stage 2 study of life-cycle emissions analysis of alternative fuels for heavy vehicles.' CSIRO, Aspendale, Victoria.

Environment Australia (2001) 'State of knowledge report: air toxics and indoor air quality.' Environment Australia, Canberra.

Frischknecht R and Jungbluth N (2004) 'Ecoinvent report no. 1 – overview and methodology.' Swiss Centre for Life Cycle Inventories, Dübendorf.

Goedkoop M and Spriensma R (1999) 'The Eco-Indicator 99. A damage oriented method for Life Cycle Impact Assessment.' PRe Consultants, Amersfoort, The Netherlands.

Grant T and Beer T (2000) 'Life-cycle assessment of alternative fuels for heavy vehicles in Australia.' *Fourth International Conference on EcoBalance*, Epochal Tsukuba, Tsukuba, Japan, 31 October–2 November 2000. The Society of Non Traditional Technology, Tokyo.

Huijbregts MAJ, Lundi S, McKone TE and van de Meent D (2003) Geographical scenario uncertainty in generic fate and exposure factors of toxic pollutants for life-cycle impact assessment. *Chemosphere* **51**, 501–508.

ISO (2001) International Organization for Standardization 'ISO 14048 – Environmental management technical specification – Life cycle assessment – Data documentation.

Lenzen M and Murray SA (2001) A modified ecological footprint method and its application to Australia. *Ecological Economics* **37**, 229–255.

NLWRA (2002) *Australian Terrestrial Biodiversity Assessment*. National Land and Water Resources Audit. Land and Water Australia, Canberra.

Olsen P, Silcocks A, Weston M and Tzaros C (2006) 'Birds of woodlands and grasslands.' Paper prepared for the 2006 Australian State of the Environment Committee. Department of the Environment and Heritage, Canberra.

Ross S and Evans D (2003) Excluding site-specific data from the LCA inventory: how this affects life cycle impact assessment. *International Journal of Life Cycle Assessment* **7**(3), 141–150.

SoE (2006) 'Australia State of the Environment 2006.' Department of the Environment and Heritage, Canberra.

Tyler M (2006) 'The disappearing frogs.' Paper prepared for the 2006 Australian State of the Environment Committee. Department of the Environment and Heritage, Canberra.

Udo de Haes HA, Jolliet O, Finnveden G, Hauschild M, Krewitt W and Müller-Wenk R (1999) Best available practice regarding impact categories and category indicators in life cycle impact assessment – background document for the Second Working Group on Life Cycle Impact Assessment of SETAC – Europe (WIA-2).' *International Journal of Life Cycle Assessment* Landsberg **4**(2), 66–74.

UN News Center (2007) 'Soaring biofuel demand driving up agricultural prices, says UN-backed report'. 5 July 2007, UN News Service, UN News Center, New York, USA. <http://www.un.org/apps/news/story.asp?NewsID=23144&Cr=biofuel&Cr1=&Kw1=food&Kw2=fuel&Kw3=>.

Wackernagel M and Rees W (1995) *Our Ecological Footprint: Reducing Human Impact on the Earth*. New Society Publishers, Philadelphia, USA.

World Wildlife Fund (2006) 'Living planet report'. World Wildlife Fund, Gland, Switzerland.

Chapter 6

Life cycle assessment and waste management

Karli L Verghese

6.1 Introduction

The consumption of products and services fuels the growth of markets and economies around the world. This growth, however, results in excessive consumption of natural resources and ever growing volumes of waste, which need to be managed (Hamilton 2003). Waste is generated from sources as diverse as residential households, offices, manufacturers, building sites, farms and mining sites. The management of household waste and the example of plastic shopping bags is the focus of this chapter.

The generation, quantity, type and subsequent management of waste are continual issues for society. Historically waste has been buried in holes in the ground – out of sight. In recent decades, the availability of land and 'holes' to bury waste has reduced, and the need to address negative environmental impacts of landfills including leachate and methane generation, and redirect valuable resources (the waste) back into the economy, have been increasingly recognised. Alternative technologies have been introduced to treat and manage waste, including reprocessing facilities, composting facilities, waste-to-energy technologies and anaerobic digestion. The overarching debate is about which technologies and processes are most appropriate, and a shift in emphasis has taken place from 'disposal' to 'management'. However, as Hayes says: 'Waste is a construct that humanity invented at a time when industry lacked a deep understanding of ecological processes. There is no waste in nature' (cited in Imhoff 2005, p. 7). Waste management practice is seen as a way to reduce or recycle waste, but in future the generation of waste may be synonymous with the generation of resources, as materials for use in an integrated closed-loop cycle.

Waste management is generally concerned with what to do with waste arising, rather than addressing the 'problem' at origin – the rising rate of materials acquisition and reducing durability of materials use. Hence, unsurprisingly, the volume of waste is rising. In Australia, the rate of waste generated per capita is among the highest in the world (OECD 2002). In 2002–03, more than 32 million tonnes of solid waste were generated by Australians, which is a rate of 4.35 kilograms per person per day (ABS 2007), most of which ends up in landfill. In Australia, most households pay a set fee for their waste and recycling services as part of their local council rates regardless of how much waste they generate. In contrast, most businesses pay for waste disposal on a fee-for-service basis. This provides the commercial sector with an incentive to minimise waste and maximise the amount of material recycled (Verghese and Lewis 2007), which may be a reason that recovery rates are lower for municipal waste than for commercial waste.

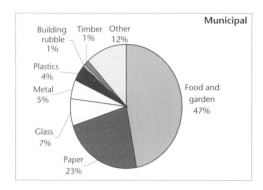
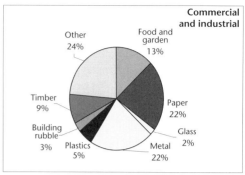
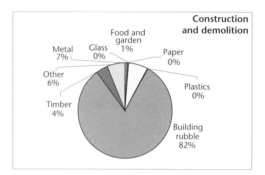

Figure 6.1   The composition of waste generated in Australia, 2002–03 (ABS 2007).

Within the literature addressing the modern concept of 'waste' and the surrounding management and policy context, LCA offers the possibility of improving our understanding of the environmental 'bottom line' of different waste strategies.

6.1.1   The waste fractions

About 42% of waste is generated in the construction and demolition sector, 29% in the commercial and industrial sector and 27% in the municipal sector (ABS 2007). An overwhelming 47% of municipal waste is food and garden waste. The rest consists of paper (23%), glass (7%), metals (5%), plastics (4%) and other (14%) (see Fig. 6.1).

It is interesting to compare Australia's waste composition with that of the USA. Although the waste categories differ slightly, only 22.2% of the USA's municipal waste stream in 2000 was food and garden waste compared with 47% in Australia. Containers and packaging, at 32.2% (75 million tonnes), was the largest waste fraction in the USA's municipal waste stream. Municipal waste represents about 55% to 65% of total waste generated in the USA (Imhoff 2005).

In the seven years from 1996 to 2003, the quantity of waste generated in Australia rose by 42%. In the same period, the quantity of waste sent to landfill reduced by 18% from 21 million to 17 million tonnes. This decline was achieved through a significant increase in recycling from 1.5 million tonnes to about 15 million tonnes (46% of waste) – a change of 879% (ABS 2006). Of the recovered waste, 53% is from construction and demolition waste, 28% is from commercial and industrial sources and 19% is from municipal waste (ABS 2007).

The recycling rate of post-consumer packaging materials in Australia (2003 baseline data) was 48%. Through the National Packaging Covenant's overarching goals, signatories are committed to a recycling target of 65% by 2010 (NPCC 2005). Refer to Table 6.1 for a comparison of recycling rates for packaging materials in Australia and the USA.

Table 6.1   Recycling percentages of packaging materials in Australia and the USA (Imhoff 2005; NPCC 2005)

Packaging materials	Australia (2003)	USA (2000)
Paper and cardboard	64%	56%
Glass	35%	26%
Steel	44%	58%
Aluminium	64%	55%
Plastics	20%	9%

When packaging waste fractions are sorted, they are sent to numerous reprocessing destinations. High-value materials such as polyethylene terephthalate (PET) and glass are generally reprocessed locally into recyclate for new packaging. In contrast, due to limited local facilities, materials such as polypropylene (PP) and mixed plastics are generally sorted together and sent to other countries for reprocessing into lower grade applications like plant pots.

When materials such as plastics and glass are sent to landfill, although they do not break down and cause leachate and emission problems, they are a lost resource which could be used to displace virgin material production in the economy. Food and garden waste, when composted and applied to soil, can increase crop yields and reduce the application of fertilisers. When organic material breaks down in the anaerobic environment of landfill, it can generate greenhouse gas emissions. LCA studies conducted in other countries to investigate the management of post-consumer waste – recyclables, organic fractions and residual wastes (e.g. Weitz *et al.* 1999; Beccali *et al.* 2001; Smith *et al.* 2001; US EPA 2002) – have demonstrated the benefits of diverting waste from landfill.

For many years governments and industry have encouraged households to recycle. By the late 1990s it was evident that there was strong public support for the separation of used packaging materials for recycling. Although this support was based on the assertion that recycling was 'good for the environment', as it saved resources and would result in less waste to landfill, scientific data was necessary to inform the debate. An Australian LCA study demonstrated that there were substantial benefits from recycling common packaging and paper materials (see Section 6.2.1). This work was further supported by the Independent Assessment of Kerbside Recycling Study, undertaken by Nolan-ITU and SKM Economics (2001) (see Section 6.2.2). However, neither study assessed management options for organic material, which is a major component of the domestic waste stream.

Over recent years there have been substantial development and diversification in technologies to deal with residual (generally not recycled) and organic waste fractions. These technologies usually provide one or more of the following outcomes:

- energy
- useful organic material
- volume reduction
- stabilisation of the organic fraction and/or toxins.

LCA can be used to indicate how any given waste management technology may affect other aspects of the waste stream and the waste management system itself, with positive or negative environmental impacts. Any evaluation of the technology needs to take the wider impacts into account. Those possible impacts were examined in a study that aimed to evaluate the environmental impacts of a range of waste management scenarios using different waste treatment (resource recovery) technologies in a full life cycle context (Grant *et al.* 2003) (see Section 6.2.3).

Another waste issue is litter, which does not lend itself easily to LCA, but which has been incorporated into LCA studies of waste. Keep Australia Beautiful has recorded that more items are being littered. Samples collected from beaches, car parks, highways, industrial areas, recreational parks, residential and retail areas and shopping centres show that 74 items per 1000 m^2 are being littered compared with 70 items per 1000 m^2 in their 2005/06 survey (McGregor Tan Research 2007). One littered item that has received particular attention in recent years is the humble high density polyethylene (HDPE) single-use shopping bag. This lightweight, although strong, material has helped many a shopper bring home their groceries, but once littered becomes an aesthetic issue when entangled on fences and in vegetation, and poses risks to aquatic and marine life of entanglement, ingestion and suffocation. In 2002, the debate around HDPE bags in Australia took a high public profile, prompted by the suggestion from Ron Clarke, a well-known athlete then representing the Council for Encouragement of Philanthropy, that a levy should be placed on shopping bags to reduce their consumption (at the time estimated to be 6.9 billion bags per year in Australia). Different types of degradable polymers were also entering the market and being touted as the 'solution' to plastic bag consumption. Streamlined LCAs of HDPE bags with alternatives such as paper and reusable bags and bags made from degradable polymers formed part of the studies commissioned by the Australian Department of Environment and Heritage (see Section 6.2.4).

6.2   Case studies in LCA application in waste policy

There have been several applications of LCA to waste management. The following case studies illustrate how LCA has been applied to municipal waste and its treatment.

6.2.1   Case study 1: the benefits of recycling paper and packaging waste

In common with LCA in many other applications, waste management studies invariably pose challenges for the collection of specific data and for the appropriate interpretation and application of findings. In a study on the kerbside collection of paper and packaging commissioned by EcoRecycle Victoria (now Sustainability Victoria) (Grant *et al.* 2001), the state department responsible for waste policy, a need was identified to collect specific Australian data on waste collection, reprocessing and management. There was public support for recycling, but no scientific data to show if there were environmental benefits. The solution was to engage with industry stakeholders through an Industry Advisory Committee, and across the three principal research groups involved: the Centre for Design at RMIT University; the Centre for Packaging, Transportation and Storage at Victoria University (part of the Co-operative Research Centre (CRC) for International Food Manufacture and Packaging Science); and the Centre for Water and Waste Technology at the University of NSW (part of the CRC for Waste Management and Pollution Control). This process ensured that any concerns or questions regarding the LCA methodology or data collection were recognised in a timely manner and dealt with appropriately. It further assisted in the collection of life cycle inventory data and in creating ownership by industry of the study's outcomes (James *et al.* 2002).

In 1998, EcoRecycle Victoria commissioned a study to investigate the life cycle impacts of domestic waste management: 'Life Cycle Assessment for Paper and Packaging Waste Management Scenarios in Victoria'. It was conducted in two stages. The first stage investigated the landfilling and recycling of glass, PET and steel (Grant *et al.* 1999). The second stage expanded the focus to investigate the entire range of paper and packaging materials presented by the consumer at the kerbside (Grant *et al.* 2001). This case study focuses on the second stage as it investigated all current packaging materials: paper and board (i.e. corrugated containers and box-board), liquidpaperboard (LPB) gable top and aseptic cartons, HDPE bottles, polyvinyl

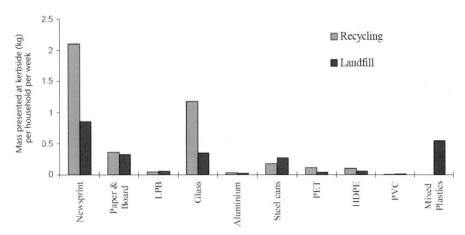

Figure 6.2 Quantity of packaging materials at kerbside and their destination. HDPE, high density polyethylene; LPB, liquidpaperboard; PET, polyethylene terephthalate; PVC, polyvinyl chloride.

chloride (PVC) bottles, other mixed plastic packaging (flexible and rigid), glass bottles and jars, steel cans and aluminium cans. It also investigated old newspapers. Survey data from 1997 showed that residents of Melbourne generated 6.6 kg of packaging and old newspaper waste per household per week, of which 4.1 kg was placed in the recycling container and 2.5 kg was placed in the garbage container (Fig. 6.2).

Landfill and mechanical recycling were, at the time, the two waste management processes in operation and were accordingly modelled (Fig. 6.3).

Modelling of the recycling system not only included a credit for reducing the negative impacts of landfill, such as virgin material avoided through recycling, and the impacts of landfill avoided through recycling, but also for positive impacts such as energy generation from landfill gas. This enabled a viable comparison with landfill. The findings of the study showed that recycling produced net savings (Fig. 6.4).

The functional unit of the study was defined as the management of the recyclable fractions of paper and board, LPB, HDPE, PVC, PET, other plastics, glass, steel and aluminium packaging, and old newspapers discarded at kerbside from the average Melbourne household in one week.

With assistance from industry representatives, this study detailed, calculated and modelled the collection, reprocessing and avoided products for all of the listed materials. The environmental savings and impacts of recycling and landfill for each material were presented in addition to the normalised values (indicator values for Australia divided per week per household). Inventory data on process systems was collected and reported, and the greenhouse gas savings and impacts from recycling each material were also reported. This provided, for the first time, a breakdown of the contribution of key activities (Fig. 6.5). For example, results showed that the benefits of collecting waste PET outweighed impacts associated with virgin PET production, even taking into account vehicles driving around the suburbs to collect recyclables, and the use of caustic soda in reprocessing and recycling processes. Similar results were achieved for the other materials.

Five environmental indicators were selected:

- greenhouse gas emissions
- smog precursors
- embodied energy
- water use
- solid waste.

Figure 6.3   The system boundary (boxed) for the paper and packaging waste project, stage 2.

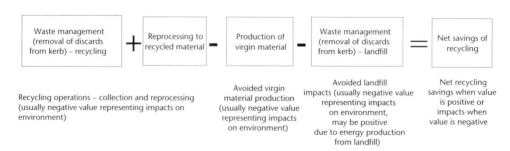

Figure 6.4   Method for calculating net environmental savings in the recycling process.

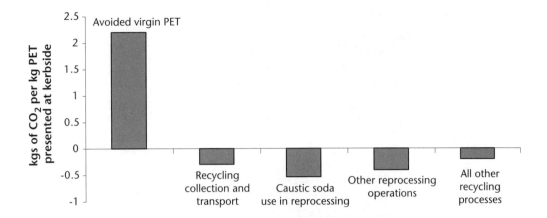

Figure 6.5 Greenhouse gas savings and impacts from polyethylene terephthalate (PET) recycling in terms of kilograms of carbon dioxide per kilogram of PET presented at kerbside.

The impacts omitted, due to limited available data or where the LCA methodology at the time was unable to handle such data, were:

- human and eco-toxicity impacts from virgin material production, recyclables and waste collection, reprocessing of materials and landfill leachate
- land use and soil impacts, particularly from forestry operations in virgin paper fibre production and agricultural operations in wheat production
- local amenity impacts from landfill
- resource consumption and/or depletion
- consumer behavioural data.

Modelling the degradation of the organic fraction took into account that carbon dioxide and methane are the by-products of degradation, with some of the methane captured for flaring or electricity generation while the remainder is lost to the atmosphere. The carbon fractions that do not degrade are sequestered and become a carbon sink (Fig. 6.6). At the time of the study, from the late 1990s to early 2000, different assumptions and approaches were used to determine the degradation rates of organic fractions in landfill. Due to great uncertainty concerning these rates and ultimate levels of degradation, three different landfill degradation scenarios were modelled:

- full degradation, in which all organic components are completely degraded
- carbon sequestration (CS US EPA data), in which 34% of newspapers and 23% of paper and cardboard is assumed not to break down. This was used as the baseline assumption, based on ICF (1997) and US EPA (1998)
- lignin content (CS lignin content calculations), in which 78% of newspapers and 53% of paper and cardboard is assumed not to break down, based on Tchbanoglous *et al.* (1993).

Depending on the degradation scenario selected, the results differ significantly for the organic fractions of old newspapers, paper and cardboard, and LPB (Fig. 6.7).

Several sensitivity analyses were undertaken:

- landfill gas capture
- reprocessing yield

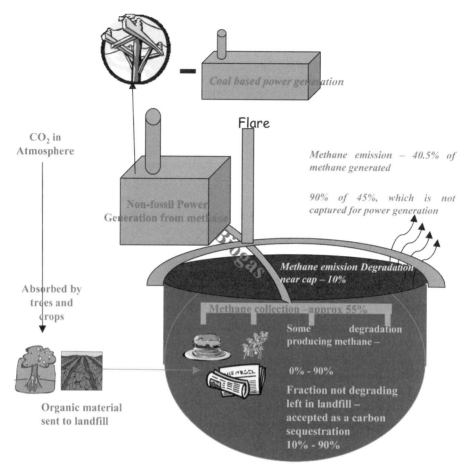

Figure 6.6   Dynamics of greenhouse gas emissions from landfill.

- yield of materials at kerbside
- household washing behaviour
- a case study on regional recycling (Bendigo)
- recycling collection frequency.

Overall, none of the sensitivities tested altered the direction of the results. However, landfill assumptions and reprocessing yields were shown to have significant potential to change the magnitude of the results.

The net savings from recycling for a typical Melbourne household per week are presented in Table 6.2 with equivalency factors used to aid interpretation.

The study concluded that the most important factors for maximising the environmental benefits from landfill are (Grant *et al.* 2001):

- recycling to the highest value product so as to avoid the production of high value, and high environmental impact, virgin materials
- maintaining or increasing the mass of materials from household catchments, without compromising the usability of the material at end-of-life – this increase in total environmental returns comes from avoided products and avoided landfill, while also making the collection more efficient on a per tonne basis

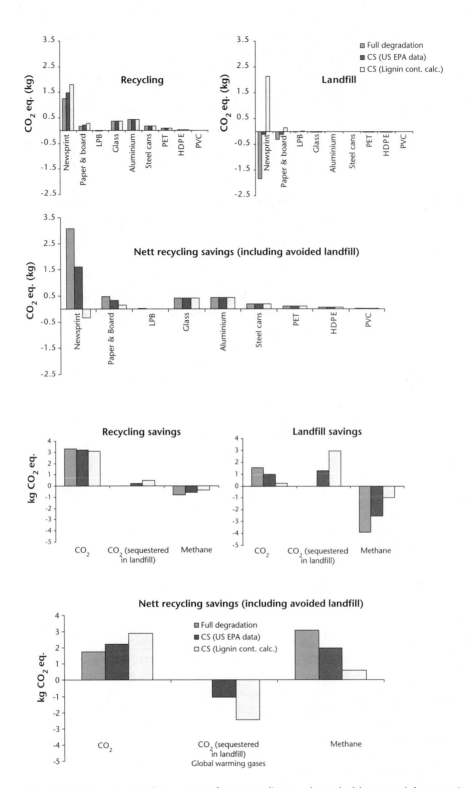

Figure 6.7 Net savings in greenhouse gases from recycling per household per week for organic degradation in three landfill scenarios.

Table 6.2   Net savings from recycling for a typical Melbourne household per week

Impact	Totals	Unit	Equivalence
Greenhouse gases	3.2	kg CO_2 eq.	This equates to 0.25% of a household's total allocation of greenhouse gases from all sources
Embodied energy	32.2	MJ	Enough energy (9 kWh) to run a 40 Watt light bulb for 72 hours (accounting for electricity losses)
Smog precursors	1.3	g C_2H_4 eq.	Equivalent to the emissions from 4.5 km of travel in an average post-1985 passenger car
Water use	92.5	litres	The equivalent of five sink-loads of dishes
Solid waste	3.6	kilogram	Depending on the material, 60% to 90% of the product put out for recycling will remain out of the solid waste stream

kg CO_2 eq., kilograms of carbon dioxide equivalents; kWh, kilowatt hours; MJ, megajoule; g C_2H_4 eq., grams of ethylene equivalents.

- reducing smog and other transport emissions from waste collection vehicles in urban areas by using efficient vehicles, with either pollution control equipment, and/or alternative fuels such as natural gas
- maintaining good landfill management practices particularly in terms of gas capture for energy recovery, landfill capping and leachate control.

It was recommended that strategies for dealing with non-recyclable paper and plastic fractions be investigated, particularly in the context of management of the broader organic material stream. Modelling was then also undertaken for five urban councils and one rural council in New South Wales, with similar findings.

6.2.2   Case study 2: environmental economics and recycling

In 1999, state and federal governments and Australian companies in the packaging supply chain signed the voluntary National Packaging Covenant, the purpose of which was to foster efficient and environmentally sustainable systems for the management of used packaging materials. In 2000, the National Packaging Covenant Council commissioned Nolan-ITU Pty Ltd and SKM Economics to undertake a study entitled 'Independent Economic Assessment of Kerbside Collection and Recycling Systems for Used Packaging Materials in Australia'. The aim of the study was to assess the net costs and benefits of kerbside collection and recycling systems and their viability (Nolan-ITU and SKM Economics 2001). It was the first time in Australia that kerbside collection and recycling of used packaging materials were examined for their financial, environmental and social costs and benefits. Previous studies had focused solely on the financial aspects. For the environmental assessment, the overarching methodology was cost-benefit analysis. This involved identifying and valuing environmental externalities of collection and recycling systems to enable the findings to be incorporated into the integrated economic assessment. Modelled inventory data was aggregated into environmental impact categories and then valued by applying environmental economic benefit assessment techniques based on published Australian government references (Nolan-ITU and SKM Economics 2001). This approach challenged the way environmental issues were evaluated and how to determine the dollar value to place upon emissions. The environmental component of the study will be the focus of this case study.

The net costs and benefits of kerbside collection and recycling systems were assessed across a range of different collection systems from 200 councils in regional and metropolitan areas in each state and territory in Australia. A range of recycling and collection systems for

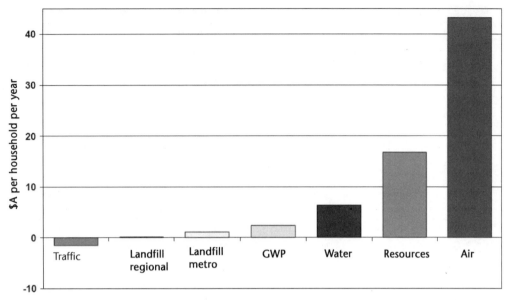

Figure 6.8   Environmental costs and benefits of kerbside recycling by impact category (A$ per household per year – population weighted national average). Savings are represented as positive values and impacts as negative values. GWP = global warming potential.

used packaging materials was compared against the baseline of all waste being sent to landfill. The systems were:

- paper and plastic collection at kerbside for recycling and energy recovery
- paper and glass collection at kerbside for recycling without energy recovery
- a shift in packaging from glass to PET
- mechanical biological treatment of domestic waste
- waste-to-energy treatment for all domestic waste.

The Integrated Solid Waste Management Model, a commercial software tool, was used to model 50 substances (resource inputs and pollutant outputs for each collection and recycling system, including the avoided product life cycle). The inventory data was aggregated into environmental impact categories and then valued through cost-benefit analysis as described above.

With variations based on waste technology and location in Australia, the national average net environmental benefit of kerbside collections and recycling systems was shown to be $68 per household per year (ranging from $41 to $119 for a collection system costing $3 per household per year). This equated to a total national environmental benefit of kerbside recycling of $424 million per year. As Figure 6.8 shows, most savings through recycling derive from the avoidance of air pollution associated with the avoided manufacture of virgin materials.

A general consensus among stakeholders, as reported by the authors, was that landfill savings are the driving force and motivation behind the recycling of materials. However, the above results, based on the value of emissions, show that landfill savings are low compared with other savings.

6.2.3   Case study 3: four waste fractions and 15 waste treatment configurations

With inventory data and impact assessment results in hand for the recycling of common packaging materials and old newspapers (as presented in Section 6.2.1), EcoRecycle Victoria commissioned a second waste study in 2002, which was completed the following year (Grant *et al.*

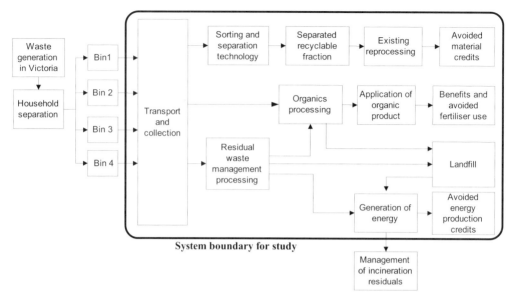

Figure 6.9   System boundary for the LCA of waste and resource recovery options (including waste-to-energy).

2003). This time, the system boundaries were expanded to investigate other kerbside waste streams and different waste treatment technologies. The findings from this study would assist in the development of Victoria's solid waste strategy: Towards Zero Waste. Activities in landfill were modelled on a more detailed level to include methane generation from degrading organic material and the creation of carbon sinks from non-degradation. This introduced a new level of complexity and insight into the role of waste generation in the economy. To reduce the greenhouse gas profile of different waste stream fractions, this study revealed some interesting findings that challenged policy makers who had the task of defining appropriate waste treatments and guiding waste management.

Since 2000, there has been substantial development and diversification of waste treatment technologies dealing with residual and organic waste fractions. Such technologies provide outcomes such as energy production, useful organic material, volume reduction and stabilisation of organic fractions and/or toxins (Grant *et al.* 2003). In light of these developments, there was a need to evaluate the overall environmental impacts of emerging waste treatment technologies in a full life cycle context so that effects on other aspects of the waste stream and waste management system could be investigated. The study aimed to provide a transparent environmental evaluation of a range of waste management technologies for dealing with mixed waste and organic waste in the Victorian waste stream.

The system boundary for the study is presented in Figure 6.9. Fifteen waste management configurations in four basic groups (A–D) were examined, presented in Table 6.3. The technology descriptions and key assumptions are presented in Table 6.4.

The environmental indicators included in the study were based on the indicator set developed by the Centre for Environmental Studies (CML) in the Netherlands (Guinee *et al.* 2001). They were:

- global warming
- resource depletion
- eutrophication
- photochemical oxidation

Table 6.3 Configurations in waste management options

Source separated organics treatment	Residual waste treatment					Paper and container recycling
	Land-fill	Aerobic stabilisation	Anaerobic digestion	Gasification/ pyrolysis	Incinera-tion	
None			A1	A2		Mixed with residual waste^A
None	B1	B2	B3	B4	B5	Separate collection of rigid containers in crate and paper via kerbside bundle
Aerobic composting green waste	C1	C2	C3	C4		
Aerobic composting green and food waste	D1	D3		D4		
Anaerobic digestion	D2					

^A Upfront sorting (metal, plastic and paper) is assumed to take place at the treatment process. Scenario A models one-stream collection, Scenario B models two-stream collection, and Scenarios C and D model three-stream collection.

- human toxicity
- eco-toxicity in freshwater, marine and terrestrial environments.

The characterisation factors for these indicators were also taken from CML with the exception of the toxicity indicators, which were taken from Huijbregts and Lundie (2002) and specifically modelled using Australian environmental conditions. Characterisation factors are used to calculate the contribution of individual substance flows in the inventory to the indicator result (i.e. for global warming, characterisation factors for carbon dioxide = 1, methane = 21 and nitrous oxide = 310).

As an illustration, Figure 6.10 presents greenhouse gas savings by gas and net savings across the 15 waste management configurations. Significant savings are achieved when green waste is diverted from landfill to composting (from scenario B1 to C1) and also food waste (scenario C1 to D1). Methane from degradation of organic waste in an anaerobic landfill environment is avoided when these waste fractions are sent to aerobic composting instead.

The key conclusions from this study were:

- Kerbside recycling delivers significant environmental benefits through avoided virgin materials and reduced energy use and can be further improved.
- Production and application of compost delivers additional benefits, including increased water-holding capacity, carbon sequestration, and reduced pesticide and fertiliser use. This study was the first to incorporate such benefits from compost applications in Australia.
- Organic residual waste treatment delivers substantial environmental benefits through avoided or reduced emissions from landfill and recovery of additional recyclable materials, particularly metals. Energy recovery from anaerobic digestion of residual waste is relatively small, after accounting for energy use in processing.
- Thermal residual waste treatments deliver greater benefits in most environmental categories. This result is based on the assumption that energy recovered from waste will replace electricity generated from south-east Australia's electricity supply system, which is largely based on black and brown coal. The issue of benefits from replacing 'dirty' electricity

Table 6.4 Technology descriptions and key assumptions for the study

Technology	Key assumptions	Description of process
Landfill of MSW	Values for methane generation from organic fractions were taken from Smith *et al.* (2001). Assume landfill gas capture is in place in landfills, accounting for 80% of overall methane generated/emitted from landfills. Of this 80%, assume 55% of methane is captured. Of this 55%, 75% results in electricity production and the remaining 25% is flared. Of the remaining 45% not captured, assume 10% degrades through the landfill cap. CO_2 is also sequestered in the landfill.	Involves direct dumping of material into clay-lined (or synthetically lined) cells where waste is compacted and covered on a daily basis with dirt. Pipes for collecting gas and leachate from the landfill are built into the waste piles as they are constructed. Collected gas may be flared or used for energy production.
Composting green waste	Recovery rates at kerbside are 90% for green waste, and 35% for other organics (non-food). 60-70% of input to process becomes compost output	Involves shredding green and food waste, placing in piles or windrows (with partial forced aeration when food waste is present), turning and refining. Benefits are achieved with application of the compost product to land.
Composting green and food waste	Recovery rates at kerbside are 65% for food 90% for green waste, and 35% for other organics; 45-55% of input to process is compost output	
Anaerobic digestion of green and food waste	Digestion is followed by aerobic compost production. Recovery rates at kerbside are 65% for food, 90% for green waste and 35% for other organics; 30–50% of input to process is compost output; 80–100 kWh/t net electricity output.	Involves processing organic material in a digester (i.e. in the absence of air), generation of biogas that is converted to electricity, and composting of digester output[A]
Aerobic stabilisation of MSW	60-70% of input to process is output. Of this output, 34% is compost and 66% is stabilised residue sent to landfill. Automated ferrous metal recovery.	Involves particle size reduction, homogenisation, composting to reduce putrescible substances, landfilling of stabilised material (which reduces volume and emission potential), recovery of metals and recovery of organic material as compost
Anaerobic digestion of MSW	55-65% of input to process is output. Of this output, 28% is low-grade compost and 72% is stabilised residue sent to landfill. Automated ferrous metal recovery; 0-20 kWh/t input net electricity output.	Involves particle size reduction, homogenisation, processing of organic material in a digester, generation of biogas which is converted to electricity, composting of digester output, landfilling of stabilised material, recovery of metals and recovery of some organic material as compost
Gasification/ pyrolysis of MSW	Output is 30% of input. Of this output, 65% is grit/slag, 30% is inert filling and 5% is residue from flue gas cleaning. Automated ferrous metal recovery, and some plastics recovery; 400-450 kWh/t net electricity generation, no steam.	A term that covers a range of new (or 'advanced') thermal waste treatment processes. The initial stages occur under restricted air (oxygen) supply, and the actual combustion of generated gas or oil (for energy recovery) occurs at a subsequent stage. The output is electricity, grit, slag, fly ash and some recyclable materials depending on the specific technology.

Technology	Key assumptions	Description of process
Incineration of MSW	Output is 32.5% of input. Bottom ash, sent to landfill, accounts for 30% of input. Fly ash, also sent to landfill, accounts for 2.5% of input; 400 kWh/t net electricity generation (no combined heat and power).	A well-proven waste processing technology overseas, extensively used in Europe, USA and Japan. Involves the combustion of waste, usually in a grate kiln, and produces electricity, slag and ash.
Recovery rates in front-end separation	20% for paper and cardboard and glass, 30% for plastics and 75% for metals	For scenarios A1 and A2 (Table 6.3), advanced front-end separation has been assumed for recovery of recyclables
Benefits of using compost from green and food waste	2.5% increase in crop yield through increase in water-holding capacity (estimated on wheat crop). Fertiliser replacement of 1.5% N and 0.25% for K and P. Nitrous oxide emission savings. 20% reduction in pesticide use. 10% of carbon sequestered in soil.	
Benefits of using compost from MSW	10% of carbon is sequestered in the soil	

[A] The digester feedstock would predominantly be food waste, with the bulkier garden waste being added at the end of the process (i.e. composting).
CO_2, carbon dioxide; K, potassium; MSW, municipal solid waste; N, nitrogen; P, phosphorus.

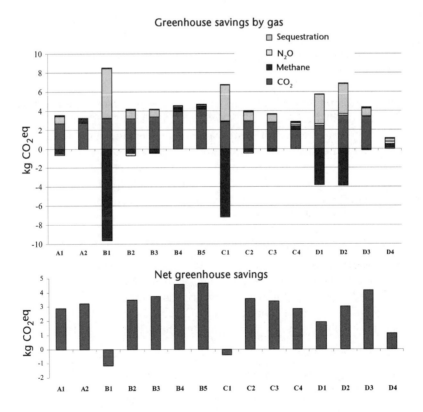

Figure 6.10   Greenhouse gas savings (kilograms of carbon dioxide equivalents) by gas and net greenhouse savings in scenarios per household per week. CO_2, carbon dioxide, N_2O, nitrous oxide.

Table 6.5 Relative contribution of waste management to energy and nutrient systems in Australia

Summary of savings compared to Australian demand[A]	Unit	Production from waste	Australian consumption per year	Percentage contribution from waste
B4 – Contribution of gasification of one year's waste to electricity	GWh	3056	211 111[a]	1.45%
D2 – Contribution of food and garden waste composting to phosphorus	Tonnes P	5560	475 500[b]	1.17%
D2 – Contribution of food and garden waste composting to nitrogen	Tonnes N	31 230	752 800[c]	4.15%

Notes (not included in the reference list in the study's report):a from ESAA (2002); b ABS (1996) based on 20% total soluble phosphorus in super phosphorus; c Brunt (2001) based on urea demand, as this is twice the size of numbers reported in ABS (1996).
[A] For an explanation of scenario codes, see Table 6.3.

affects the environmental indicators of resource depletion, and human, terrestrial, marine and freshwater aquatic toxicity. Again, some environmental impacts from these technologies, such as toxicity from residue disposal, could not be incorporated.

The thermal recovery options were also incompatible with collection and composting of source-separated organics, due to competition for the organic fraction as a feedstock for both these processes.

In evaluating the tension between composting and energy pathways for organic material, it is important to examine the broader system dynamics, and establish the extent to which waste contributes to overall problems associated with an energy project and overall loss of nutrients from Australian farmland.

Table 6.5 shows the respective contributions of the best energy and composting options to Australia's overall demand for energy and nutrients. For energy recovery, waste has the potential to supply 1.4% of Australia's electricity demand. In nutrient terms, the best composting option would supply more than 4% of Australia's nitrogen demand and about 1% of our phosphorus demand. While neither nutrient contribution is great enough to significantly affect the markets in which they operate, if waste composting was combined with recycling of sewage solids and other organic material, it could close a nutrient cycle in Australia's poor soils.

The results of the study have been compared with similar quantitative studies of waste management options (e.g. US EPA 2002) and the European Commission (e.g. Smith *et al.* 2001). This comparison recognises that the end results of life cycle-based studies differ according to the study goals, assumptions and methodological choices. All the studies reviewed highlight the importance of these choices in regard to the end results. In general, all studies indicate improved environmental performance associated with dry material recycling. There is an overall benefit, and most environmental categories are improved, from source separated aerobic or anaerobic management of organic waste. Also in general, landfill performs the worst of all technologies in relation to resource depletion, photochemical oxidation, water toxicity and greenhouse gas potential.

6.2.4 Case study 4: plastic fantastic?

Humble lightweight HDPE shopping bags are produced in large numbers and have rapidly become ubiquitous – and invariably free – at supermarket checkouts. The combination of light weight and high strength has ensured their popularity, and it is estimated that 6.9 billion bags are used per year in Australia. However, their widespread use, single-use disposability

Table 6.6   Composition of all shopping bags modelled including assumptions

Bag material	Composition	Assumptions made
Degradable polymers		
Starch polybutylene succinate/ adipate (PBS/A) (e.g. Bionelle)	50% – starch from maize; 25% – 1,4-butanediol; 12.5% – succinic acid; 12.5% – adipic acid	Adipic acid is manufactured from cyclohexane (40%) and nitric acid (60%); succinic acid is formed through fermentation of corn-derived glucose
Starch with polybutylene adipate terephthalate (PBAT) (e.g. Ecoflex)	50% – starch from maize; 25% – 1,4-butanediol; 12.5% – adipic acid; 12.5% – terephthalate acid	1,4-butanediol is derived either from natural gas or corn glucose
Starch-polyester blend (e.g. Mater-Bi)	50% – starch from maize; 50% – polycaprolactone (PCL)	Maize-growing data is based on data from the Netherlands. PCL is produced from cyclohexanone (95%) and acetic acid (5%).
Starch-polyethylene blend (e.g. Earthstrength)	30% – starch from cassava (tapioca); 70% – high-density polyethylene	Cassava-growing data is based on cassava-growing data from the Netherlands
Polyethylene + prodegradant (e.g. TDPA)	97% – high density polyethylene; 3% – additive	Additive was modelled as stearic acid and a small amount of cobalt metal to represent the presence of cobalt stearate
Polylactic acid (PLA)	100% polylactic acid	Based on maize growing in USA
Alternatives		
Singlet HDPE	HDPE	Production of HDPE film
Kraft paper bag with handle	Kraft virgin pulp	Production of paper bags
PP fibre 'green bag'	PP	Production of PP film
Woven HDPE 'swag bag'	HDPE	Production of HDPE film
Calico	Cotton	Cotton processing
LDPE 'bag for life'	LDPE	Production of LDPE film

LDPE, low density polyethylene; HDPE, high density polyethylene; PBAT, starch with polybutylene adipate terephthalate; PBS/A, starch polybutylene succinate/adipate; PCL, polycaprolactone; PLA, polylactic acid; PP, polypropylene; TDPA, total degradable polymer additive.

and litter potential have contributed to increasing scrutiny of the environmental impacts associated with their manufacture, use and disposal. Bags that are littered often become entangled on fences, in trees or in waterways where they can threaten aquatic life and interfere with the visual aesthetic of the natural environment. Alternative materials such as calico, paper and woven polypropylene have been introduced as alternatives to 'single-use' HDPE. Degradable polymers have been marketed more recently as a possible solution that would reduce the demand for a non-renewable resource (HDPE) by replacing it with a biodegradable renewable resource (e.g. maize-derived), thereby potentially reducing littering problems and demand upon landfills.

In the study reported here (Scheirs *et al.* 2003), a streamlined LCA of a selection of degradable plastics suitable for applications in film blowing and marketed as materials for shopping bags were compared with HDPE, LDPE, PP, Kraft paper and calico (Table 6.6).

The function of the study was defined as the use of shopping bags to carry groceries and goods from store to home. The number of single-use bags required and the number of reusable bags required to carry goods home per person per year were calculated. The functional unit was defined as a household carrying home about 70 grocery items from a supermarket each week for 52 weeks (Table 6.7).

In order to understand how degradable plastics degrade in aerobic and anaerobic environments, four different end-of-life waste management treatment technologies were modelled for all the different bags (Table 6.8):

- landfill (anaerobic environment) – modelled on Victorian landfills
- source-separated green and food MBT (mechanical biological treatment) composting
- municipal solid waste MBT composting
- municipal solid waste anaerobic digestion.

As the littering of HDPE bags results in aesthetic issues and endangers aquatic and marine life through entanglement and ingestion, the following two litter scenarios were developed, and each bag was modelled against them (Table 6.9):

- litter aesthetics (calculated based on the time the bag would be litter (metre squared per year (m^2y))
- litter's effects on marine biodiversity (calculated based on whether the polymer floats and how long it floats, or whether it sinks and how long it takes to sink).

Figure 6.11 illustrates the greenhouse gas emission profile for the six degradable polymers compared with six alternative materials: two single-use materials (i.e. HDPE and Kraft paper) and four reusable materials (i.e. calico, PP fibre 'green bag', woven HDPE 'swag bag' and LDPE

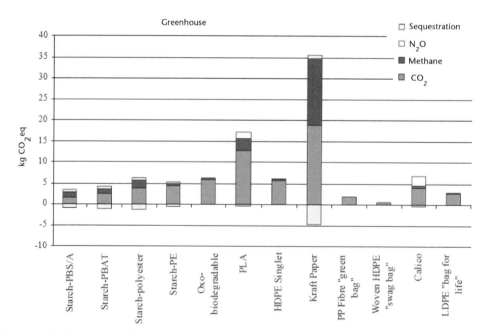

Figure 6.11   Breakdown of greenhouse gas emissions for all bags. CO_2, carbon dioxide; N_2O, nitrous oxide. PBS/A, starch polybutylene succinate/adipate; PBAT, starch with polybutylene adipate terephthalate; PE, polyethylene; PLA, polylactic acid; HDPE, high density polyethylene; PP, polypropylene; LDPE, low density polyethylene.

Table 6.7　Characteristics of the use of all bags

Degradable polymers	Weight (g)[A]	Relative capacity	Quantity of bags per week in relation to relative capacity	Expected life	Quantity of bags per year adjusted in relation to expected life	Transport to Australia
				Use of bags		
(Starch-PBS/A)	8.1	1 (6–8 items)	10	Single trip	520	From Italy (16 000 km)
(Starch-PBAT)	6	1 (6–8 items)	10	Single trip	520	From Japan (8000 km)
(Starch-polyester)	6	1 (6–8 items)	10	Single trip	520	50% from Germany (16 000 km) and 50% from USA (13 000 km)
(Starch-PE)	6	1 (6–8 items)	10	Single trip	520	From Malaysia (6000 km)
(oxo-biodegradable)	6	1 (6–8 items)	10	Single trip	520	Concentrate from Canada (16 000 km) and 50% of bags from Malaysia (6000 km)
PLA	8.1	1 (6–8 items)	10	Single trip	520	50% from USA (13 000 km) and 50% from Japan (8000 km)
Singlet HDPE	6	1 (6–8 items)	10	Single trip	520	Hong Kong (7000 km)
Kraft paper bag with handle	42.6	1	10	Single trip	520	N/A
PP fibre 'green bag'	PP 65.6 Nylon base 50.3	1.2	8.3	104 trips (2 years)	4.15	N/A
Woven HDPE 'swag bag'	130.7	3	3.3	104 trips (2 years)	1.65	Taiwan (7000 km)
LDPE 'bag for life'	40	2	–	10 trips (1 year)	26	N/A
Calico bag	125.4	1.1	9.1	52 trips (1 year)	9.1	Pakistan (11 000 km)

[A] Based on the mass required to perform the same function as an HDPE singlet bag
LDPE, low density polyethylene; HDPE, high density polyethylene; N/A, not applicable; PE, polyethylene; PLA, polylactic acid; PP, polypropylene; PBS/A, starch polybutylene succinate/adipate; PBAT, starch with polybutylene adipate terephthalate.

Table 6.8 End-of-life assumptions for alternative bags

Alternative bags	Landfill %	Recycled %	Composting %	Litter %	Reuse (as a bin liner for household waste)%[A]
All degradable polymers	70.5	0	10	0.5	19
HDPE singlet bag	78.5	2	0	0.5	19
Kraft paper bag with handle	39.5	60	0	0.5	0
PP fibre 'green bag'	99.5	0	0	0.5	0
Woven HDPE 'swag bag'	99.5	0	0	0.5	0
LDPE 'bag for life'	97.5	2	0	0.5	0
Calico bag	99.5	0	0	0.5	0

LDPE, low density polyethylene; HDPE, high density polyethylene; PP, polypropylene
[A] Subsequently avoids HDPE bin liners. In the landfill environment and source-separated organics composting it is assumed that degradable polymers will degrade like food waste (i.e. 90% of the polymer will degrade).

'bag for life'). The findings for greenhouse gas emissions indicate that reusable bags, except for calico, still have a lower impact upon the environment than HDPE singlet bags or degradable polymers. Greenhouse impacts are dominated by carbon dioxide through electricity and fuels consumption, methane emissions through degradation of materials in anaerobic conditions (e.g. landfill), and nitrous oxide (N_2O) emissions in fertiliser applications on crops (Fig. 6.11). Degradable polymers with starch content produce more greenhouse gas emissions due to methane emissions during landfill degradation and N_2O emissions from fertilising crops. A sensitivity analysis was performed to determine the effect of reusing Kraft paper bags. In the results in Fig. 6.11 it is assumed that the Kraft paper bags are single trip bags. In the sensitivity it is assumed that each bag will be reused a second time (i.e. that only 260 bags are required in

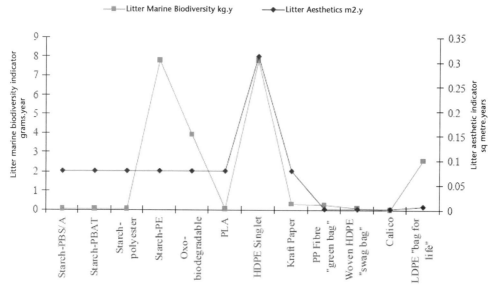

Figure 6.12 Litter characterisation values for all bags. The values chosen here have been estimated in the absence of definitive data on the subject and are presented to show how the potential marine impact may vary under these assumptions. PBS/A, starch polybutylene succinate/adipate; PBAT, starch with polybutylene adipate terephthalate; PE, polyethylene; PLA, polylactic acid; HDPE, high density polyethylene; PP, polypropylene; LDPE, low density polyethylene.

Table 6.9   Characteristics of all bags in different littering environments

Alternative bags	Litter by area	Litter by mass	Litter: aesthetics	Litter: marine biodiversity
Mater-Bi	30 × 20 cm = 0.06 m²a	8.1 g	Assume bag litter lasts for 6 months[A] (0.03 m²y)	Sinks in 1 day (0.0221 g/year)
Bionelle	30 × 20 cm = 0.06 m²a	6 g	Assume bag litter lasts for 6 months (0.03 m²y)	Sinks in 1 day (0.016 g/year)
EcoFlex	30 × 20 cm = 0.06 m²a	6 g	Assume bag litter lasts for 6 months (0.03 m²y)	Sinks in 1 day (0.016 g/year)
Earthstrength	30 × 20 cm = 0.06 m²a	6 g	Assume bag litter lasts for 6 months (0.03 m²y)	Floats for 6 months (3 g/year)
Oxo-biodegradable (PE and prodegradant additive)	30 × 20 cm = 0.06 m²a	6 g	Assume bag litter lasts for 6 months (0.03 m²y)	Floats for 3 months (due to prodegradant) (1.5 g/year)
PLA	30 × 20 cm = 0.06 m²a	8.1 g	Assume bag litter lasts for 6 months (0.03 m²y)	Sinks in 1 day (0.016 g/year)
HDPE singlet bag	30 × 20 cm = 0.06 m²a	6 g	Assume bag litter lasts for 2 years due to light film (0.12 m²y)	Will float for 6 months (3.5 g/year)
Kraft paper bag with handle	20 × 30 cm = 0.06 m²a	42.5 g	Assume bag litter lasts for 6 months (0.03 m²y)	Assumed to sink in 1 day (0.116 g/year)
PP fibre 'green bag'	42 × 42 cm = 0.09 m²a	115.9 g	Assume bag litter lasts for 5 years (0.45 m²y)	Assumed to float for 6 months (58 g/year)
Woven HDPE 'swag bag'	50 × 50 cm = 0.18 m²a	130.7 g	Assume bag litter lasts for 5 years (0.9 m²y)	Assume to float for 6 months (65 g/year)
LDPE 'bag for life'	42 × 42 cm = 0.09 m²a	125.4 g	Assume bag litter lasts for 2 years (0.18 m²y)	Assumed to float for 6 months (62 g/year)
Calico bag	42 × 42 cm = 0.09 m²a	125.4 g	Assume bag litter lasts for 2 years (0.18 m²y)	Sinks in 1 day (0.34 g/year)

m²a, metre squared area; m²y, metre square per year; LDPE, low density polyethylene; HDPE, high density polyethylene; PE, polyethylene; PLA, polylactic acid; PP, polypropylene.

A This is an assumption. There is limited data available on degradation rates in litter. Estimated 'shelf-life' is six months for TDPA-EPI plastics, and two years for starch – polyester (Hall 2003). Estimates of degradation times depend on both the resin and the environment, and range from two months to more than a year.

one year compared with 520 bags in the single-use scenario). Clearly, the impacts of the Kraft paper bag are halved if the bag is used twice by the consumer.

The single-use bags produce more litter due to the higher incidence of them being littered compared with reusable bags (Fig. 6.12). The marine biodiversity category is mostly affected by the propensity of the material to float or sink. Higher impacts are modelled in the marine biodiversity category if the material floats, as it is assumed to float for six months (three months for the oxo-biodegradable bag). If it sinks, the material is assumed to sink over the course of one day.

Polymer-based reusable bags have less environmental impacts than all of the single-use bags. Degradable bags have similar greenhouse gas and eutrophication impacts to conventional HDPE bags. If the degradable material can be kept out of landfill and managed through composting, the greenhouse gas impacts will be reduced but not eliminated. The synthetic polymer bags have greater effects on resources (abiotic depletion). In the study, indicators for litter are developed in an attempt to represent some of the damage effects caused by litter. Litter impacts are lowest for the reusable bags, but of all the single-use bags, the biodegradable ones generally produce fewer emissions, although in the marine environment, it is the density of the material that dominates the impact rather than degradability.

6.3   Conclusions

The first three case studies all indicate that dry material recycling is associated with improved environmental performance. There is an overall benefit from source-separated aerobic or anaerobic management of organic waste, for which most environmental categories show an improvement over conventional landfill. Generally, landfill performs the worst of all technologies in relation to resource depletion, photochemical oxidation, water toxicity and greenhouse gas emissions.

All of the case studies presented indicate the importance of carefully designed goal and scope, assumptions and methodology in waste LCA. There is significant potential for changes in assumptions to affect end results; for instance, recovery rates of materials in the recycling system or the degradation rates of organic materials in anaerobic environments. Some form of sensitivity analysis is important wherever there is significant uncertainty in key assumptions.

Methodological and technical considerations aside, the application of LCA has been successful in significantly influencing waste management policy in Australia. For example, in relation to the first case study, the equivalency factors generated have been used by Sustainability Victoria (formally EcoRecycle Victoria) to report the benefits of recycling in the Annual Survey of Victorian Recycling Industries reports. The second case study provided input into the National Packaging Covenant, and the third case study on alternative waste treatment technologies was used in the development of Victoria's solid waste strategy: Towards Zero Waste.

The application of LCA has also shown that recycling generally makes good environmental sense, while also challenging the notion that biodegradable polymer or paper-based materials are 'better' than petroleum-based plastics. Indeed, in the case of shopping bags only multi-use bags are likely to outperform 'traditional' lightweight HDPE bags. Nevertheless, there remain significant environmental problems associated with the generation and management of waste, and LCA can be expected to continue to assist in identifying both intuitive and counter-intuitive performance aspects of different waste questions over the coming years and decades.

6.4   References

ABS (1996) *Australians and the Environment*. Australian Bureau of Statistics, Commonwealth of Australia, Canberra.

ABS (2006) *4613.0 Australia's Environment Issues and Trends 2006.* Australian Bureau of Statistics, Commonwealth of Australia, Canberra.

ABS (2007) *1301.0 – Year Book Australia.* Australian Bureau of Statistics, Commonwealth of Australia, Canberra.

Beccali G, Cellura M and Mistretta M (2001) Managing municipal solid waste: energetic and environmental comparison among different management options. *International Journal of Life Cycle Assessment* **6**(4), 243–249.

Brunt JV (2001) *Dampier Nitrogen Project.* Retrieved 25 May 2004 from <http://www.plentyr.com.au/>.

ESAA (2002) *Electricity Australia 2002.* Electricity Supply Association of Australia, Melbourne.

Grant T, James K, Dimova C, Sonneveld K, Tabor A and Lundie S (1999) 'Stage 1 report for life cycle assessment of packaging waste management in Victoria.' Stage 1 of the National Project on Life Cycle Assessment of Waste Management Systems for Domestic Paper and Packaging. Report to EcoRecycle Victoria, Melbourne. Centre for Design at RMIT University, the Centre for Packaging Transportation and Storage at Victoria University (as part of the CRC for International Food Manufacture and Packaging Science) and the Centre for Water and Waste Technology at the University of New South Wales (as part of the CRC for Waste Management and Pollution Control), Melbourne and Sydney.

Grant T, James K, Lundie S and Sonneveld K (2001) 'Stage 2 report for life cycle assessment for paper and packaging waste management scenarios in Victoria.' Report to EcoRecycle Victoria, Melbourne. Centre for Design at RMIT University, Centre for Packaging Transportation and Storage at Victoria University and the Centre for Water and Waste Technology at University of New South Wales, Melbourne and Sydney.

Grant T, James K and Partl H (2003) 'Life cycle assessment of waste and resource recovery options (including energy from waste).' Report to EcoRecycle Victoria. Centre for Design at RMIT University and Nolan-ITU Pty Ltd, Melbourne.

Guinee JBFE, Gorree M, Heijungs R, Huppes G, Kleijn R, de Koning A, van Oers L, Wegener Sleeswijk A, Suh S, Udo de Haes HA, de Bruijn D, van Duin R, Huijbregts M, Lindeijer E, Roorda AAH, van der Ven BL and Weidema BP (2001) *Life Cycle Assessment. An Operational Guide to the ISO Standards.* Ministry of Housing, Spatial Planning and the Environment (VROM) and Centre for Environmental Science, Leiden University (CML), The Netherlands.

Hall W (2003) Personal communication between Warwick Hall, Plastral Fidene (agents for Mater-Bi) and John Schiers of ExcelPlas Australia.

Hamilton C (2003) *Growth Fetish.* Allen and Unwin, Crows Nest, Sydney.

Huijbregts M and Lundie S (2002) 'Australian life cycle impact assessment of toxic emissions.' Department of Environmental Studies, Nijmegen University (The Netherlands) and Centre for Water and Waste Technology, UNSW (Australia).

ICF (1997) 'Greenhouse gas emissions from municipal waste management – draft working paper.' Prepared for Office of Solid Waste and Office of Policy, Planning and Evaluation, US Environmental Protection Agency, Washington DC.

Imhoff D (2005) *Paper or Plastic. Searching for Solutions to an Overpackaged World.* Sierra Club Books, San Francisco.

James K, Grant T and Sonneveld K (2002) Stakeholder involvement in Australian paper and packaging waste management LCA study. *International Journal of Life Cycle Assessment* **7**(3), 151–157.

McGregor Tan Research (2007) 'Keep Australia Beautiful National Litter Index, Annual Report 2006/2007.' Frewville, South Australia.

Nolan-ITU and SKM Economics (2001) 'Independent assessment of kerbside recycling in Australia.' Revised Final Report – Volume 1 for the National Packaging Covenant Council. Manly, New South Wales, Nolan-ITU Pty Ltd and Sinclair Knight Merz, Sydney.

NPCC (2005) *The National Packaging Covenant – Strategic Partnerships in Packaging. A Commitment to the Sustainable Manufacture, Use and Recovery of Packaging*, 15 July 2005 to 30 June 2010. National Packaging Covenant Council, Melbourne,

OECD (2002) *Environmental Data Compendium*. Organisation for Economic Co-operation and Development, Paris.

Scheirs J, Lewis H, Grant T, James K, Allen P and Lenihan V (2003) 'The impacts of degradable plastic bags in Australia.' Final Report to Department of the Environment and Heritage. ExcelPlas Australia, Centre for Design at RMIT University and Nolan-ITU, Melbourne.

Smith A, Brown K, Bates J, Ogilvie S and Rushton K (2001) 'Waste management options and climate change.' Final report to the European Commission, DG Environment, AEA Technology, Abingdon, UK.

Tchbanoglous G, Theisen H and Vigil S (1993) *Integrated Solid Waste Management*. McGraw-Hill, Singapore.

US EPA (1998) 'Greenhouse gas emissions from management of selected materials in municipal solid waste.' Final Report for the US Environmental Protection Agency under EPA Contract No. 68-W6-0029, US Environmental Protection Agency, Washington DC.

US EPA (2002) *Solid Waste Management and Greenhouse Gases. A Life-Cycle Assessment of Emissions and Sinks*. US Environmental Protection Agency, Washington DC.

Verghese K and Lewis H (2007) Environmental innovation in industrial packaging: a supply chain approach. *International Journal of Production Research* **45**(18/19), 4381–4401.

Weitz K, Barlaz M, Ranjithan R, Brill D, Thorneloe S and Ham R (1999) Life cycle management of municipal solid waste. *International Journal of Life Cycle Assessment* **4**(4), 195–201.

Chapter 7

Life cycle assessment: applications in the built environment

Ralph E Horne

7.1 Introduction

The built environment is where most Australians spend the majority of their time. This includes homes, commercial, retail and leisure developments, and the transport and services infrastructure to support these facilities. Here, the main emphasis will be on buildings, since this is mostly where life cycle assessment (LCA) has been applied in the built environment to date.

Buildings are a significant component of the human environment and, accordingly, contribute both to the economy and environmental impacts, including global climate change. They also present a 'classic' LCA problem, since they consume considerable amounts of material and energy (and therefore create impacts) during at least two major life cycle phases: construction and occupation. There is therefore a longstanding debate over the optimal balance between the following strategies:

- minimising material use
- investing in more materials to create a more energy-efficient building, which reduces impacts through the occupation phase.

This issue will be discussed later in the chapter. Another significant issue concerns the maintenance and alteration of buildings. The built environment is continually being renovated, refurbished and added to, providing considerable opportunities to use LCA in optimising interventions and specifications. Indeed, in housing alone, reinvestment in existing housing is about A\$16 billion per annum in Australia (in 2005), which is the same order as investment in new housing construction, yet reinvestment incurs relatively little environmental regulation. In addition, Australians have been adding to the total building stock at 3.8% (A\$35.5 billion) per annum in recent years, and in the three years to 2004/05, spent \$77.7 billion per annum on a combination of new buildings and alterations and additions to existing buildings: that is, 9.3% of Gross Domestic Product (DEWR 2007). Across OECD countries including Australia, buildings consume 30% to 50% of available raw materials and account for 25% to 40% of final energy consumption, while generating about 40% of total waste to landfill (OECD 2002, 2003).

7.1.1 Towards sustainable built environments

Globally, most of the built environment is located in cities, and there are challenges to be overcome before cities can progress significantly towards sustainability (Girardet 2004).

Research in Australia indicates that Australian cities are no exception and that change is necessary in the way we plan, configure and live in our suburbs (e.g. Newman and Kenworthy 1999; Lowe 2005). Challenges arise in policy, political, institutional, infrastructural, social, cultural and environmental terms, throughout the various planning, design, construction, maintenance, occupation, renovation and recycling/end-of-life phases of buildings. One of the most significant environmental challenges is Australia's high greenhouse gas emissions per capita. This presents two significant problems. First, in order to meet global community obligations to mitigate climate change, these current high emissions must be reduced significantly. Second, high energy users are more vulnerable to a future where greenhouse gas emissions will have a high cost, and where sustainable energy resources will be the norm. In order to build resilience, Australian communities can anticipate and adapt to climate change by reducing their need for fossil fuel-based energy resources.

The concept of 'greenhouse-neutral' (or 'carbon-neutral') buildings and communities has attracted considerable attention in recent years. A 'greenhouse-neutral' building can be defined as a structure where greenhouse gas emissions associated with the combustion of fossil fuels (crude oil and derivatives, coal, lignite, natural gas and other fossilised organic remains in shales and related petroleum deposits) are equivalent to a net of zero. It is important to distinguish between 'greenhouse-neutral' and 'zero emission'. A greenhouse-neutral building in the near future is still likely to produce greenhouse gas emissions although, almost inevitably, to a lesser extent than is currently the case. However, it will be greenhouse-neutral in that it offsets these emissions by a variety of means such as the production and export of electricity from non-greenhouse gas emitting sources, sequestration of carbon emissions by tree planting or geo-sequestration, or through carbon trading.

Of course, greenhouse gas emissions are a paramount issue, but these are not the only sources of environmental impact caused by the built environment. Further progress in reduction of both greenhouse gases and other environmental impacts will inevitably require new policies, regulation and assessment rigour.

7.1.2   Policy and regulation

LCA can assist both with policy and regulation formulation (e.g. through macro-level studies), and with compliance (e.g. through the provision of case-specific information in standardised, comparative terms). Policy and regulation to improve environmental performance of the built environment is in its infancy – internationally and in Australia. The main area which has so far received attention has been operational heating and cooling efficiency. A previous policy vacuum at federal level is partly addressed through the implementation of the '5 Star' standards, providing nationwide energy efficiency building regulations for all classes of new buildings from mid-2007. These have their origins in state-level standards introduced in Victoria in 2004, and the incorporation of initial nationwide energy efficiency provisions into the Building Code of Australia in 2003. The vehicle for the code is the Australian Building Codes Board, which has its origins in a state agreement to support a common code in 1994.

However, despite the lead-in and history to 5 Star, the introduction of these new regulations has not occurred without considerable resistance – and collaboration (Horne *et al.* 2007). Within this debate, questions raised include whether the new regulations set standards too high (or too low); what economic costs may be involved; and whether the tools, timescales and institutional arrangements are appropriate. An Australian study that compared 5 Star standard housing with standard housing being built in the United States of America (USA), Canada and the United Kingdom (UK) indicated that the latter is performing at the equivalent of almost seven stars on average. This means that Australian residential buildings are still producing 30% to 40% more greenhouse gas emissions from operational heating and cooling requirements than equivalent western world housing (Horne and Hayles 2008). This study also indicated that mandatory building codes are effective mechanisms for setting benchmark standards for aspects of building environmental performance. Meanwhile, the issue of regulating for embodied energy or end-of-life has yet to be taken up significantly by policy makers.

In contrast, the voluntary regulatory and performance arena is marked by a relative plethora of rating tools, guidelines and checklists for ecologically sustainable development (ESD). Although rating tools invariably started with attempts to improve energy efficiency, more recently, tool development has shifted into the broader arena of environmental sustainability performance. The National Australian Built Environment Rating System (NABERS) and the New South Wales Building Sustainability Index (BASIX) identify water use and stormwater as major issues in addition to energy and greenhouse gases, and the Green Building Council of Australia's Green Star suite of tools identifies a range of environmental performance issues to be considered. A next logical step may be for regulation to follow into this wider area of environmental performance. However, many of the 'sustainability' rating tools are still in their infancy. Any regulation must be built upon a robust evidence base and tools used in this should therefore be:

- based on underlying science
- comprehensive in terms of the impact scope assessed
- usable and practical
- transparent in methodology and algorithms
- internally consistent and valid
- suited to climate, cultural and other factors in the area to which they apply
- not overly resource-intensive in operation
- clear in stepping up performance levels over time to those required by long-term goals
- reviewed and updated regularly to reflect developments in knowledge.

LCA-based tools have existed since the 1990s, developed with buildings assessment applications in mind or specifically to address buildings environmental assessment or design issues. For example, the Boustead model has been extensively used for building assessment, and computer-aided design (CAD)-related tools have been developed, such as BEES, Building Design Advisor (USA) and, more recently, LCA Design (Australia). Athena and Optimise were both developed in Canada, to provide support in designing buildings with the environment in mind; there are numerous other examples. Ecospecifier was developed in Australia as a guide to environmentally sustainable and healthy products, materials and technologies for the construction sector, specifically targeted to the needs of decision-makers and specifiers. However, like the main rating tools (e.g. LEED in the USA, BREEAM in the UK and Green Star in Australia), Ecospecifier borrows from LCA but does not maximise its use.

Since LCA fulfils the requirements listed above, there is a *prima facie* case for the increased use of LCA in contributing to regulation and policy for environmental performance of buildings, underpinned by international standards. In particular, LCA provides a scientific basis for the way in which environmental impacts are assessed and combined, and provides rigorous and transparent information on these impacts. For example, it can be used to estimate the total energy consumption of buildings and even the specific fossil-based energy component. In turn, this assists in specifying the life cycle profile – in this case, of greenhouse gas emissions. It can specify 'life cycle total fossil fuel consumption' (as opposed to simply 'energy efficiency') and other environmental impacts of materials and indoor environmental quality (e.g. indoor climate, air and daylight). If these impacts are to be regulated, LCA will be important. The following sections illustrate the use of LCA in a range of built environment applications over recent years, current applications and prospects for future applications.

7.2 Case studies in LCA application in the built environment

As introduced above, the built environment is a relatively long-lived component of the economy and can therefore be seen as an asset. However, as buildings consume energy during operation – notably in maintaining internal comfort levels, operating lighting and appliances, and in maintenance and refurbishment, they also generate environmental impacts. More thermal mass and insulation, and better design using ESD principles, can reduce the operational energy burden. However, this may lead to more energy used in producing and constructing buildings in the first place. There is therefore a question about the optimal balance between the use of products and materials in buildings, and a buildings' efficiency during operation: a question well-suited to LCA, given its suitability for comparing different life cycle stages of a product or service.

There have been numerous applications of LCA to the problem of embodied energy versus operational energy. The first case study (see Section 7.2.1) illustrates a comparison of embodied and operational energy use in a major public building. The second case study (see Section 7.2.2) takes a Year 2055 time horizon in recognition that buildings are more long-lived than other 'products', and also extends the limits of traditional LCA application at the 'per unit of function' level, by investigating total building materials flows and environmental impacts in Australia. The study achieves this by combining LCA with other methodological approaches in a novel way. The third case study (see Section 7.2.3) also adopts LCA and extends the traditional approach, this time recognising the performance of building materials in assemblages. For example, rather than simply assessing bricks as an external cladding material, the Building Assemblies and Materials Scorecard approach allows the life cycle performance of the whole wall assembly to be quickly and accurately assessed. The scorecard approach reflects that buildings are generally more bespoke or unique than other products, and so a specific method of evaluation on a 'per product' basis is useful. The fourth case study (see Section 7.2.4) examines the application of LCA in a buildings regulation setting in the Netherlands.

7.2.1 Case study 1: Stadium Australia study

Stadium Australia, the centre piece of the Sydney Olympics site, is a well-known example of an Australian building where LCA was actively used in the design and construction process. The stadium was built to host the 2000 Olympic Games in Australia and is intended to continue to be used to host sporting events and concerts for over 50 years to come. The LCA was carried out by the New South Wales Department of Public Works and Services and ERM Mitchell McCotter (DPWS 1998). The results were used in detail – to inform the design process; and in summary – as documentation to verify that the planning policy had been met, to demonstrate responsibility to the various environmental interest groups, and finally to document the 'green' Olympics as an example for others to follow. This case study draws on the original studies of Janssen and colleagues, and a previous case study by Hes (2003).

The planning policy for Olympic Games projects No. 38 (State Environment Planning Policy 38) required compliance with ESD and the Environmental Guidelines for the Summer Olympic Games (Sydney Olympics 2000 Bid Environment Committee 1993), specifically that the project consider environmental impacts over its life cycle (i.e. in its manufacture, use and disposal). This led to the builder, Multiplex Construction, using LCA to quantify the stadium's environmental performance. In one of the company's papers, LCA is described as a method that:

> can be used to quantify and assess the environmental impact of any product, system or service. It is well suited to describing the environmental impact of buildings and their associated services … LCA provides a way to quantify the relative importance of building use compared to the rest of the building's life cycle. (Janssen 1998)

The goal of the LCA was to quantify the impacts of Stadium Australia throughout its life in order to minimise those impacts. Specifically, the objective was to quantify raw material use, energy use, emissions to air and water, and solid wastes, and to create an inventory of results (Building Innovation and Construction Technology 1999). The scope of the study included (Janssen 1999):

- procurement (raw materials extraction, manufacture and transport) of the building systems
- construction and reconfiguration
- operation and maintenance for a 50-year design life
- demolition.

As the LCA was used to quantify the impacts of the current design for optimisation purposes, the functional unit was the provision of a stadium for 50 years. The functional unit was further split into different life cycle stages (Matthew Janssen to Dominique Hes, pers. comm., 21 February 2001), summarised below:

- Total life cycle – The functional unit for the total life cycle was the sum of the functional units for procurement, construction and reconfiguration, operation and maintenance and demolition stages.
- Procurement – The functional unit for the procurement stage was the raw materials extraction, processing and transport of the major building materials to the stadium site.
- Construction and reconfiguration – The functional unit for this stage was the construction of the stadium (i.e. the building systems considered in procurement) and the reconfiguration of the stadium post-Olympics.
- Operation and maintenance – The functional unit for this stage was the operation and any major maintenance required of the stadium over its 50-year design life to accommodate up to 110 000 spectators during the Olympics and up to 80 000 spectators post-Olympics. The results of this stage of the stadium's life cycle were directly dependent on the forecast use of the stadium in terms of numbers of events and spectators per year.
- Demolition – The functional unit for this stage was the demolition of the stadium at the end of its design life of 50 years.

The project considered the impacts of the materials (including transport), the use of the materials, the use/maintenance of the building, and the disposal of the building and their associated emissions, without aggregating them. Building system procurement processes considered included the extraction of raw materials, manufacture of the products and systems, and transport. All major processes required to procure the building systems such as the production of energy and intermediate transport, were also included. The subsystems were aggregated as shown in Table 7.1.

Soft furnishings were excluded from the study, as were any systems outside the boundary of the stadium, such as the precinct works and outbuildings. In all cases, systems excluded were considered minor in relation to the overall works. Similar breakdowns of information were used to describe the stadium's construction and reconfiguration, operation and maintenance, and demolition. The data was collected using a quantitative questionnaire, with assistance provided to help in its completion. Building product suppliers were contacted for a description of their manufacturing processes and associated raw materials, energy use, water use and waste products. The disadvantage of this approach was that much of the data was collected during procurement after many of the design decisions had been made. Data was also collected from

Table 7.1 Subsystem aggregations

Concrete systems	Ceiling and wall systems
Concrete in situ	Masonry block walls
Pre-cast concrete	Cement render
Bored piles	Plasterboard, fibreboard
Retaining wall	Tiling
Steelwork systems	Glasswork
Hand rails, barriers gates and other steelwork	Interior paintwork
Facade metal cladding	**Seating and roof systems**
Structural steelwork	Stadium seating
Building services systems	Polycarbonate roof
Hydraulics system	**Other systems**
Mechanical and air conditioning system	Lifts and escalators
Electrical system	Arena track
Fire service system	
Stormwater system	

other studies especially for the operational phase; for example, forecast energy use for the building (Rudds 1998) and forecast water use (Sinclair Knight Merz 1998) (cited in Janssen and Buckland 2000).

7.2.1.1 Results

Table 7.2 summarises the LCA results for energy use, greenhouse gas emissions, solid waste production and water use (DPWS 1998, cited in Janssen 1999).

At the procurement stage, concrete and steelwork dominated. Most energy was used in making the materials and only 6% was used in transport. However, as shown in both Table 7.2 and Figure 7.1a, the procurement energy impact was found to be less significant (18%) than the operational energy impact of the building over the 50-year estimated life cycle. The LCA showed that an estimated 675 000 tonnes of solid wastes would be produced over the life of the building, 385 000 tonnes of which would arise during demolition, and would therefore comprise of mixed building materials. The assumed waste routes are shown in Figure 7.1b.

In terms of comparative performance, the LCA showed that, in its operational phase, the stadium used 30% less energy than other conventional stadiums with the same functional unit (Fig. 7.2a). Further reductions were quantified in water, with up to 77% of total water used either sourced from recycled water or collected on-site (Fig. 7.2b).

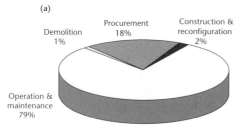
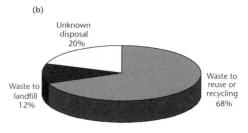

Figure 7.1 (a) Energy and (b) waste performance (after Janssen 1999).

Table 7.2 Summary of Stadium Australia LCA results (Janssen 1999)

	Procurement	Construction and re-configuration	Operation and maintenance	Demolition	Total
Primary energy use (TJ)	1370 18%	150 2%	6000 79%	80 1%	7600
Greenhouse gas emissions ('000 tonnes CO$_2$ eq.)	140 22%	10 2%	470 75%	5 1%	625
Solid wastes – recycled or landfilled ('000 tonnes)	80 12%	50 7%	160 24%	385 57%	675
Water use ('000 tonnes)	680 22%	90 3%	2250 74%	5 >> 1%	3025

TJ, teraJoules (10^{12} Joules); '000 tonnes CO$_2$ eq., thousand tonnes of carbon dioxide equivalent greenhouse gas emissions.

7.2.1.2 Discussion

LCA was used in this project primarily to help Multiplex Construction to quantify the environmental performance of Stadium Australia; to set a benchmark for future stadiums and to ensure that the regulatory and planning requirements were met. To expedite the collection of data, the requirement for data collection was integrated into the contractual arrangements. According to Buckland, it is the first time subcontractors have participated in an inventory, and 'collection of information was not easy' (Building Innovation and Construction Technology 1999). Despite the contractual requirements, the timing of the data collection was problematic and provided a challenge for the timely completion of the LCA. On the positive side, this innovative approach paid off in this early example of the use of LCA in assisting the builder in choosing environmentally preferable building materials, and also helping inform design issues such as waste and energy efficiency (Janssen and Buckland 2000).

　　While there are many lessons from this case study, in hindsight, perhaps the most significant is how little LCA activity occurred in building and construction in the years immediately

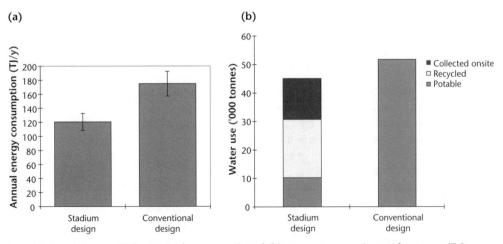

(a)　　　　　　　　　　　　　　　　　**(b)**

Figure 7.2 (a) Energy (TJ/y, teraJoules per year) and (b) water comparative performance (7.2a after Janssen and Buckland 2000; 7.2b after Janssen 1999).

following the Olympics project. Without a clear incentive for designers or builders to carry on such work, the use of LCA in this type of application was extremely limited until the Green Building Council Green Star and other building rating tools became prominent, and demand for environmental information on building materials rose once again. This is unfortunate as some of the tools developed in the early to mid-1990s, such as LCAid, were not further developed. Had their development continued, this would have placed LCA in a strong position to meet current demands for building LCA modelling. Instead, they withered for want of a market.

7.2.2   Case study 2: scoping impacts of building materials in Australia

The federal Department of the Environment and Heritage commissioned this study in 2005 to provide information to the Australian Building Codes Board following the inclusion of 'sustainability' in the inter-governmental agreement that underpins the Building Code of Australia. This case study is drawn from the resulting report (DEWR 2007). The research aimed to identify and quantify the range of environmental impacts associated with the building fabric and to identify possible measures that could improve the sustainability of building materials across the life cycle or supply chain.

The study examined the life cycles of key materials that make up the building fabric. The choice of materials affects operational performance, although the link is not always precise or direct. For example, two buildings with apparently identical materials can have dramatically different operational energy performance depending on design, detailing and construction. The buildings themselves and fit-out materials, including operational aspects of building use such as lighting, space heating or cooling and use of appliances, were excluded from the study. Infrastructure construction such as roads, bridges, services and materials used outside the building (e.g. paving, driveways and fencing) was also excluded.

Materials flows for the Australian construction sector were established using historical data collected by BIS Shrapnel over the years 1994 to 2005 for four sectors: new separate houses, improvements to separate houses, multi-residential construction and non-residential construction. This data was used in combination with the CSIRO Australian Stocks and Flows Framework (ASFF), a population-driven materials flow model for the Australian economy, to provide the base case scenario, which was set at 2005. The ASFF was used to project likely trends in the use of building materials for the next 50 years (until 2055), extrapolated from historical trends over the last 20 years using the materials flows identified. The resulting materials flow data was then assessed with SimaPro software using inventory data relevant to Australia (with input from the Australian building products sector) to estimate the potential environmental impacts occurring from production, use and disposal of materials – now and into the future.

A business-as-usual model, with unabated growth in house demand, was developed for comparison with a range of scenarios including the base case to test the relative impacts of different strategies to reduce environmental impacts. The total list of scenarios were:

*   unabated growth in house sizes (business-as-usual)
*   base case (set at 2005)
*   increased shift to multi-unit residential properties
*   reduction in demolition rate of houses
*   slow lowering of house size
*   increased number of people per household
*   fast lowering of house size
*   combined reduction in housing size and increase in persons per household.

Where clear trends were not observable in the historical period, scenario values were fixed at the 2005 value. The average size of new houses has grown significantly, by about 40%, over the past 20 years to a current average of 258 m^2. The study assumed that average new house size will continue to increase slightly and then decline to 240 m^2 by 2055, and that multi-unit residential properties will stabilise at 115 m^2 over the same period. A projected growth in building demand over the next 50 years was also taken into account. No major changes in relative percentages of materials used (e.g. through shifts in market preference) were assumed within each building type.

Stakeholder consultation about the research findings, including the provision of expert advice and additional contacts with industry bodies, assisted in developing the range of measures which formed the recommendations of the study. Building industry associations and individual companies provided life cycle data that enriched and expanded the existing data sets held by the Centre for Design at RMIT University, while interviews with industry practitioners explored barriers to the uptake of more sustainable products and practices. Stakeholder workshops were held in Sydney, Melbourne, Brisbane, Adelaide, Perth and Hobart during which a wide range of comments, questions and ideas were received from the 160 workshop participants.

7.2.2.1 *Results*

The quantities of materials used in buildings and their environmental impacts have risen sharply in the 20 years to 2005. The largest flows identified occur in the new residential construction sector followed closely by the home improvement sector, then by non-residential construction and multi-residential construction. As a consequence of the growth in house size, even though material intensity per square metre has decreased significantly, this material efficiency has been offset by faster growth in building size.

Building materials included in the study were responsible for the generation of about 2% of total Australian greenhouse gas emissions and 10% of the overall greenhouse gas emissions of buildings. Concrete, steel, aluminium and brick are typically the most significant contributors to construction impacts (Fig. 7.3).

At the assemblage level, greenhouse gas emissions can vary up to 400% between lower and higher performers, with some generating as much as 10 times more emissions than others. Also, different activities produce different patterns of impacts. Figure 7.4 presents a summary of the contribution to LCA indicators by each construction type as indicated by the SimaPro analysis.

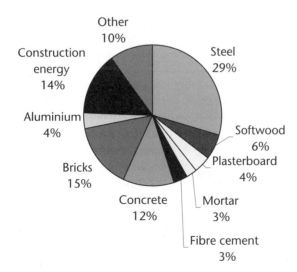

Figure 7.3 Greenhouse gas emissions by material for all building sectors for 2005 (DEWR 2007).

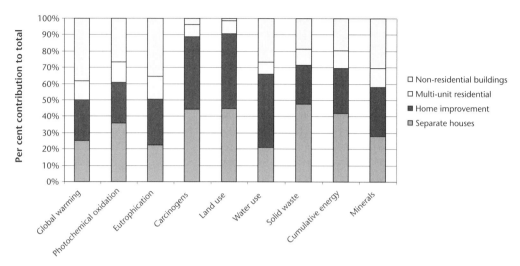

Figure 7.4   Contribution to indicators by each construction type (DEWR 2007) ('Global warming' is interchangeable with greenhouse gas emissions).

Assuming new house size will decline slightly and multi-unit residential properties will stabilise by 2055, total materials use by mass will grow by almost 40% due to growth in building demand. As a consequence, greenhouse gas emissions will rise by 40% and water use by 63%. Concrete use is expected to more than double over the next 50 years, with bricks making up a decreasing share partly due to the shift to multi-unit residences and away from separate houses. The remaining materials of significance are steel, softwood, plasterboard and mortar.

Modelling of the eight future scenarios showed that a combination of options, incorporating existing or currently forecast technology, has the potential to reduce greenhouse gas emissions from materials use by 45% by 2055 from the unabated growth (business-as-usual) scenario (i.e. to hold impacts at 2005 levels). Figure 7.5 illustrates the results for each scenario,

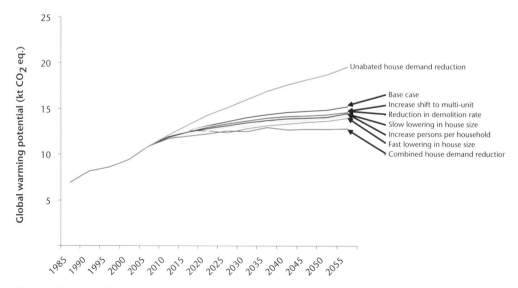

Figure 7.5   Annual greenhouse impacts from building materials from 1985 to 2055 under different building demand scenarios (kilotonnes of carbon dioxide equivalents) (DEWR 2007).

and shows how greenhouse gas emissions would not be reduced overall without either stringent measures beyond those modelled and/or the coupling of these measures with others (e.g. a major switch to renewable electricity generation away from fossil fuels). This analysis excludes efficiencies achieved through optimisation at design and construction stages. These and other approaches would therefore be required in order to achieve a net reduction in total impacts from 2005 levels.

7.2.2.2   Discussion

The use of LCA clearly illustrates the potential relative 'savings' of different strategies. Among these strategies, a reduction in the number and size of buildings (or projected new buildings) provides the most significant potential greenhouse gas savings. The study also indicates that a range of interventions is required if the rise in buildings-related greenhouse gas emissions is to be arrested – and then reduced. Since much of the environmental impact of buildings is determined at the design stage, through materials decisions, in-built efficiencies and performance criteria incorporated into the design, the next area for impact reduction is improvements in building design. This was beyond the scope of the current study and is undoubtedly a fertile area for further research. For today's building designers and specifiers, it can be difficult and costly to obtain credible environmental performance information, and some form of widely recognised labelling would assist where there is demand for environmentally preferable building materials. However, additional measures to support application of ESD and sustainable building materials, such as additional building code requirements, are also important in driving uptake.

The study found that the environmental costs of increasing primary industry and raw materials extraction, where this is required to improve buildings and materials environmental performance, are likely to be small and insignificant compared to the benefits obtained over the building life cycle. However, significant life cycle impacts occur at end-of-life, either through direct impacts (emissions and outputs to the environment, and energy inputs to processes) or through the lost recycling opportunity and associated additional requirements for new materials. About 60% of construction and demolition waste is recycled. The rate of recovery tends to be higher for metals and other high value materials, but many materials that are recycled, such as concrete, are 'down-cycled' into roadbase and other low value uses. Economically, there is significant potential benefit from increasing resource recovery. Preliminary analysis suggests that recovering an additional 5% to 10% of the value of building materials currently resulting from demolition (i.e. 5% to 10% of A$1.5 billion, or A$75–150 million per annum) may be a reasonable target.

A further option to reduce greenhouse gas emissions and other environmental pollution and resource impacts is to optimise materials individually through process, manufacturing and supply-chain innovation. This includes improving product durability. Specification for best life cycle performance within given assemblies can be achieved through the provision of credible tools and information applicable to a wide range of settings, with appropriate drivers (regulatory and/or voluntary) to assist uptake.

Finally, designing for the life of buildings – alterations, 'churn' (refits) and design for disassembly – and applying such activities in efficiency-maximising ways has considerable potential to reduce impacts, while also providing economic benefits to building owners, facility managers and tenants over the building's life. For example, using an 'open building' system allows alterations in the building layout without significant construction work. This involves separating the structure from the cladding and providing access to all parts of the building and all components, which can help to minimise impacts through maintenance. Similarly, using components that are sized to suit the intended means of handling can reduce waste during materials use at all stages.

In preparing for effective and efficient disassembly, the building should be designed in such a way as to use a minimum number of different types of connectors while providing a means of handling components, with realistic tolerances to allow for disassembly activities. Also, permanent identification of component types should be included, and a hierarchy of disassembly should be envisaged, related to the expected life span of components.

Presently there is no systematic reporting or measurement framework for construction materials in Australia. LCA is sometimes undertaken and some governments have set up guidelines for ensuring that best practice measures are adopted. However, a systematic and consistent approach to building materials would be valuable in informing ongoing policy development and identification of significant opportunities. The need for development and use of tools allowing easy, accurate and quick quantification of environmental costs and benefits of design is a key potential mechanism for change towards more sustainable buildings, when introduced in concert with other (e.g. regulatory, market or pricing) measures. The following factors would all contribute to this endeavour:

- improved information flow
- expanded product stewardship
- environmental labelling
- design standards for occupant productivity and well-being
- measures to facilitate materials recycling and component reprocessing.

LCA constitutes an appropriate technique to deliver on these requirements.

7.2.3   Case study 3: BAMS – a building assemblage LCA design tool

The Building Assemblies and Materials Scorecard (BAMS) is a logical extension of buildings LCA and macro-level assessment. It involves the development of quick reference information on the environmental performance of building assemblages. The project is supported by the Sustainability Fund managed by Sustainability Victoria. Project partners include the Green Building Council Australia, VicUrban, The City of Port Phillip, Moreland and Manningham city councils, and the Victorian Department of Sustainability and Environment.

The main drivers behind the development of BAMS are:

- a growing understanding of environmental impacts of building products and materials, and identification of the potential for dramatic improvements in Australia and other areas
- an indication that through green design and construction practices the use of building materials is potentially an important solution rather than a problem as it has been in the past.

At its core, BAMS is a standardised method and reporting format. Building upon initiatives such as the Green Guide to Specification in the UK (Anderson 2002) and Milieu Relevante Product Informatie (MRPI) (Environmentally Relevant Product Information) in the Netherlands (MRPI 2007), BAMS uses LCA to ensure consistent and science-based evaluation of construction options. The potential for this type of decision-support tool spans an array of applications, including providing a common basis for materials assessment in environmental building rating-tools. BAMS scorecards could become invaluable to project teams aiming to optimise total building performance over buildings' lives, and to suppliers of building products seeking to differentiate high-performing products and assemblies from others.

BAMS provides a step change in LCA application to buildings in Australia by:

- automating and speeding up project-specific LCA data and assessment information provision, enabling it to be used through the design phase

- basing information on scientific life cycle data rather than qualitative estimates
- incorporating links and support for the Australian National Life Cycle Inventory
- considering the whole-of-life performance, including end-of-life disposal or reuse
- examining environmental performance based on the application – including where and how building products are used in the building (e.g. wall, floor or roof)
- bringing together key tool-developing organisations to jointly develop and implement a common methodology, offering for the first time a single reporting format for materials performance assessment
- supporting manufacturers, suppliers and design teams in responding pro-actively to growing interest in the environmental performance of building products and materials.

As the first point above indicates, in providing readily accessible LCA information produced using a standard protocol and in a standard, scorecard format, BAMS provides readily comparable data on building-assemblage environmental performance for design teams wishing to incorporate life cycle performance aspects of building elements into their early decision-making. The development of standard criteria for assessing product sustainability provides both a common basis and transparency for both manufacturers and specifiers. This in turn provides signals for future innovation and investment. Hitherto, invariably, only generic information in the form of generalised LCA data and ESD principles has been available. BAMS provides the ability to quantify the benefits of improved approaches at all stages in the supply chain, and for project design teams to make evaluations on a project-specific basis.

BAMS is an ongoing research and development initiative. The initial project, completed in 2008, provided a methodological framework and scoring of several generic (non-branded product) assemblies. Potential further developments include the application of accreditation and verification systems, refinement of LCA data, and expansion of generic assemblies. There is also potential for incorporating brand-specific data through the expansion of BAMS datasets, and integration with the National Life Cycle Inventory.

7.2.4   Case study 4: built environment LCA – the Dutch experience

There is no regulation, existing or planned, to mandate application of LCA in Australian buildings. In considering potential future mechanisms, it is useful to consider the Dutch experience for comparative purposes, where more regulation has occurred. The background drivers in the Netherlands were growing concerns for environmental protection from government and some manufacturers in the building industry in the 1980s and 1990s. Of significance was the 1989 formation of MBB (Milieuberaad Bouw, an environmental 'think tank' for the building industry). Following this, government and the building industry worked together to develop environmental policies for the building sector. A successful project led to increased reuse and recycling of construction and demolition waste from 60% in 1990 to 90% in 2000.

In 1995 the MBB group suggested the consistent application of LCA in the building industry, by which time decision-makers at council level were often using materials preference lists developed using qualitative criteria based on notions of environmental preferability (e.g. no tropical hardwood, no PVC, 'recyclable', 'recycled content'). Bossink (2002) summarised contemporary Dutch practice around the turn of the millennium:

> Most of the methodologies are being developed at universities and at research
> institutes related to universities (Bijen and Schuurmans 1994) and used by
> consultants in the construction industry to develop practical lists with
> environmentally friendly design options (Anink and Mak 1993; Haas 1994;
> Stofberg 1995; Stofberg *et al.* 1996). Frequently, these lists are used in
> demonstration projects.

The 1995 MBB initiative was aimed at bringing increased rigour to such processes and 'mainstreaming' environmental performance across all relevant products, manufacturers, practices and trades. Also in 1995, the Energy Performance Coefficient (EPC) was introduced, whereby newly built houses were required to incorporate energy efficiency measures to meet EPC requirements. These energy performance standards are now updated every few years. In 1996, the Dutch government in cooperation with more than 14 building associations initiated the development of standardised sustainable construction options incorporating environmental criteria (Jansen 1996). In 1998, the environment was added as the fifth pillar of the principles underpinning housing regulation.

By the late 1990s, the development of methodologies for LCA of materials, energy and buildings was being actively pursued by the Dutch government. LCA was chosen as a means of providing a level playing field while adopting widely accepted, rigorous methodologies and science, and promoting full functionality-based comparison of options across all building life cycle stages. The vehicle chosen was the MRPI system, which has its origins in 1997 when it was first commissioned by the Nederlands Verbond Toelevering Bouw (NVTB), the association of Dutch building products suppliers. By mid-1999, the first edition of the MRPI manual was complete and the first manufacturers were having their products assessed and reviewed. In late November 1999, the MRPI was formally introduced and 30 MRPI certificates were presented.

Soon after the introduction of MRPI, the MRPI Foundation secured cooperation with the Stichting Experimentele Volkshuisvesting (SEV, the steering committee for experiments in public housing) and Stichting Bouwresearch (SBR, the foundation for building research) who are the initiators of Eco-Quantum, a tool that calculates the environmental performance of buildings (IVAM 2007). In 2000, MRPI and Eco-Quantum became compatible, enabling MRPI results to be entered directly into Eco-Quantum. As manufacturers produced new or updated MRPI certificates, their information could be added to the Eco-Quantum database. However, the process was slow and during 2001 to 2002, development was also slow in both Eco-Quantum and the addition of MRPI certificates to the database. In 2002, Eco-Quantum version 2 was introduced with a simplified interface for users.

Concurrently, other tools became established, including Greencalc, used particularly by local governments, and the Materiaalgebonden Milieuprofiel van Gebouwen (MMG, 'material-related environmental profiles for buildings'). MMG was developed by Nederlands Normalisatie Instituut (NEN, the standards institute for the Netherlands), using LCA methodology and building on experience from the MRPI and Eco-Quantum. For many practitioners, the MMG initiative marked a logical extension of LCA into the regulatory context, since the aim was to introduce a performance requirement into the building code and thus mandate the use of MMGs. However, this initiative failed in 2003, when it was withdrawn due to lack of support at the policy level. The project was left with a longer term goal of applying MRPI in a European context (e.g. by the International Organization for Standardization Technical Committee TC59/SC 17, *Sustainability in Building Construction*).

It is timely and relevant to reflect briefly on the Dutch experience with the application of LCA to building materials, especially the attempt to set standards and develop mandatory regulations based on LCA-derived specifications for the environmental performance of building materials. Undoubtedly, institutional and political factors in the Netherlands had a significant role in both the planning and early adoption of the LCA-based strategy, and in its subsequent demise. These factors vary from place to place and are not the main consideration here. Suffice to say that the support of a range of individuals and agencies, organised in a strategic manner, would be an essential precondition for LCA to have a successful regulatory role in Australia.

There were also clear technical reasons that the strategy ran into problems. These included classic LCA challenges, such as the technique requiring many assumptions and choices to be made during modelling, and data may not always be available. To overcome this, the Dutch system included a penalty model designed to encourage the industry to deliver complete and correct data. Incomplete data led to default conditions applying, which typically led to overestimation of results by 20% or more. However, this approach to developing and expanding the reference database made the LCA complex and time-consuming. It also ultimately interfered with the transparency of the process and the results, and had the effect of raising the costs and risks of the LCA work.

In summary, the technical system that was developed in order to overcome lack of data ultimately led to a significant and ultimately insurmountable burden falling on the process, given the balance of institutional and political factors operating at the time. It would be a mistake to conclude from this example that regulation and LCA do not mix well, or that LCA cannot be applied successfully in a mandatory compliance framework. Rather, the key lesson is that data development needs to be in place from the outset, and the necessary processes to update data must be designed not to inhibit regulation's key functions, which include transparency and ease-of-use.

7.3   Directions for LCA in the built environment

The built environment assists human societies to meet basic needs for shelter and security. Increasingly, it is also being developed to provide high levels of comfort and amenity, with considerable environmental impacts. LCA has a major role in highlighting the impacts of different built environments across their various life cycles. Five directions can be identified for LCA to assist improving the sustainability of the built environment over the next decade:

1. Increasing attention on reducing greenhouse gas emissions through improving the thermal efficiency of buildings is driving changes in building design. Some of these changes lead to a corresponding increase in embodied energy loads. There is therefore a potential 'optimal' investment of embodied energy for operational energy payback in some cases. LCA has an ongoing role in improving understanding of such issues and in providing support for design decisions. This will become more pressing as operational energy efficiency improves, since the proportion of total energy-related impacts attributable to embodied energy will rise. Since many buildings and their siting are unique, there is the potential to further develop life cycle management tools to assist in this process.

2. There is a plethora of building rating tools. Their emphases vary, and many incorporate some 'life cycle thinking', although most do not incorporate LCA to any significant extent. With improvements in LCA methods and building data over the past decade or so, and with the increasing share of embodied energy in the total energy impact, there is a strong case for the increased use of LCA in many building rating tools. This use may vary from embodied energy calculators or indicators to the development of more sophisticated multi-impact assessment-based tools to support design decisions. LCA can also assist in researching and setting appropriate functional units for tools; for example, is 'energy use per house', 'energy use per bed space', or 'energy use per square metre' an appropriate unit, and what is the sensitivity of results to assumed maintenance regimes and design life?

3. There are environmental impacts associated with buildings that normally fall outside building regulations. These include, for example, energy consumed by products used in but not part of the building fabric, such as appliances. Here, LCA can provide common metrics and therefore 'bridge' the area between consumer labelling and buildings' environmental performance.

4. LCA has a role in urban design, where spatial disciplines and scale-based tools are commonly used. While LCA typically assumes a functional unit, and therefore does not readily focus

on scale, it can contribute importantly in assessing the total environmental impacts of different design options. This is important in an era where there is likely to be a range of novel so-called solutions to climate change at a range of urban scales, from distributed energy and water systems to higher density urban forms (which have higher embodied energy and generally use different construction materials and techniques).

5. The development of LCA information on building materials on an assemblage basis is a rapidly developing picture, placing Australia at the forefront of research in this area. Moreover, the lessons from the Dutch experience are useful in informing the requirements and needs of a system to support the wider uptake of LCA-based building materials information.

Finally, as practices for making additions to the built environment become more sustainable, increasing attention is being paid to the greatest part of the problem: the existing stock. This invariably represents a range of forms built over at least the last 150 years with 'bursts' of additions in the late 19th century and in the inter-war (1930s) and post-Second World War (1950s) periods. Arguably, much of this stock requires retro-fitting and renovating urgently to improve its environmental performance. There is a need to apply LCA to assist in deciding when and how to update buildings appropriately for optimum environmental outcomes. Again, this issue mainly concerns a balance of embodied and operational energy, but also encompasses obsolescence, durability, use-efficiency and related issues that apply to new buildings.

7.4   References

Anderson JDS (2002) *The Green Guide to Specification*. 3rd edn. Blackwell Science.

Anink D and Mak J (1993) *Manual Sustuinable Heuse Building (Handleiding Duurzame Woningbouw)*. Steering Group for Experiments in Residential Building (SEV): Rotterdam (in Dutch).

Bijen JM and Schuurmans A (1994) *MBB-Project Environmental Measures in Construction (MBB-Project Milieumaten in de Bouw)*. Milieuberaad Bouw (MBB), Sittard (in Dutch).

Bossink BAG (2002) A Dutch public-private strategy for innovation in sustainable construction. *Construction Management and Economics* **20**(7), 633–642.

Building Innovation and Construction Technology (1999) 'Theory In To Facts' section, Number 5 February 1999. Retrieved 6 February 2001 from <http://www.dbce.csiro.au/inno-web/0299/greenolympics.htm>.

DEWR (2007) Scoping Study to Investigate Measures for Improving the Environmental Sustainability of Building Materials. DEWR.

DPWS (1998) NSW Department of Public Works and Services: Stadium Australia Life Cycle Assessment (Inventory Results). Conducted for Multiplex Constructions (NSW) by DPWS Environmental and Energy Services and ERM Mitchell McCotter.

Girardet H (2004) *Cities People Planet*. Wiley-Academy, Chichester.

Haas M (1994) *Environmental Classijication of Building Materials (Milieuclassificatie Bouwmaterialen)*. Dutch Institute for Biological and Ecological Construction (NIBE): Bussum (in Dutch).

Hes D (2003) Unpublished paper on Stadium Australia for Greening the Building Life Cycle, Life Cycle Assessment Tools in Building and Construction. Course by Centre for Design, RMIT University, Melbourne.

Horne RE and Hayles C (2008) Towards global benchmarking for sustainable homes: An international comparison of the energy performance of housing. *Journal of Housing and the Built Environment* **23**, 119–130.

Horne RE, Dalton T and Wakefield R (2007) Greening housing in Australia: a question of institutional capacity. *ENHR 2007 International Conference 'Sustainable Urban Areas'*. Rotterdam, The Netherlands, European Network for Housing Research.

Horne RE, Opray L and Grant T (2006) Integrating Life Cycle Assessment into housing environmental performance. In: *Proceedings of the 5th Australian Life Cycle Assessment Society Conference*. Melbourne. Australian Life Cycle Assessment Society (ALCAS), Melbourne.

IVAM (2007) *Eco-Quantum Description*. <http://www.ivam.uva.nl/uk/producten/product7.htm>.

Jansen PFC (1996) *Sustainable Construction: National Package for House Building (Duurzaam Bouwen: Nationaal Pakket Woningbouw)*. Foundation for Building Research (SBR), Rotterdam (in Dutch).

Janssen M (1998) Life Cycle Assessment of Buildings and Services. *AIRAH International Conference*. Sydney.

Janssen M (1999) 'Stadium Australia life cycle assessment.' Cited in *Greening the Building Life Cycle*. Retrieved 3 January 2009 from <http://buildlca.rmit.edu.au>.

Janssen M and Buckland B (2000) Stadium Australia Life Cycle Assessment – Energy Use and Energy Efficient Design. Cited in *Greening the Building Life Cycle*. Retrieved 3 January 2009 from <http://buildlca.rmit.edu.au>.

Lowe I (2005) *The Big Fix*. Black Inc, Melbourne.

MRPI (2007) *Milieu Relevante Product Informatie*. Retrieved 26 March 2008 from <http://www.mrpi.nl>.

Newman P and Kenworthy J (1999) *Sustainability and Cities: Overcoming Automobile Dependence*. Island Press, Washington DC.

OECD (2002) *Design of Sustainable Building Policies: Scope for Improvement and Barriers*. OECD, Paris.

OECD (2003) *Environmentally Sustainable Buildings: Challenges and Policies*. OECD, Paris.

Stofberg FE (1995) *Environmenal Checklist New Estates of Houses (Milieuchecklist Nieuwbouwwoningen)*. Office for Environmental Research and Design (BOOM): Delft (in Dutch).

Stofberg FE, Van Hal A, Matton T and Kaiser MA (1996) *Building Blocks for a Sustainable City (Bouwstenen voor een Duurzame Stedenbouw)*. Association of Dutch Municipalities (VNG): The Hague (in Dutch).

Sydney Olympics 2000 Bid Environment Committee (1993) Environmental Guidelines for the Summer Olympic Games, Sydney Olympics 2000 Bid Ltd, Sydney.

Chapter 8

Will the well run dry? Developments in water resource planning and impact assessment

Andrew S Carre and Ralph E Horne

8.1 Introduction

Humankind's dependence on water is universally recognised, although our planning, management and use of it have not always sufficiently valued it as a finite resource. Indeed, as Adam Smith pointed out some time ago: 'Nothing is more useful than water: but it will purchase scarce nothing; scarce anything can be had in exchange for it' (Smith 1776). Treating this unique, limited and clearly useful resource as virtually 'free' inevitably results in problems in maintaining sustainable supply along the lines of Garret Hardin's *Tragedy of the commons* (Hardin 1968). In Australia, rationing has long been practiced through allocation systems, notably in the vast Murray-Darling basin. However, perennial questions arise as to the appropriateness of allocations in terms of both size and proportions, to different activities and regions.

Moreover, recent developments such as the widespread recognition that anthropogenic global climate change is exacerbating normal drought cycles have led many communities to reassess the value of water and the way it is managed and used. Australia is a noteworthy global example in this regard, since climate and land management pressures have combined to make the driest inhabited landmass (Smith 1998) even drier and many areas have experienced severe drought and the prospect of catastrophic water failure.

Phillip Island, near Melbourne, is but one example of a region with acute water shortages that have threatened collapse of the reticulated water system: Westernport Water chief David Mawer has warned Bass Coast Council that Phillip Island could run dry if residents failed to curb consumption. 'That would be the extreme case, but we have a reservoir that is not that big, so if it doesn't rain, then yes, we will run out' (Houston 2006).

In this brave new water-scarce world, life cycle assessment (LCA) has emerged as a tool in quantifying and understanding the causes of water shortage and assessing mitigating strategies. By applying LCA, users and suppliers of water are revealing inefficiencies in existing water practices and designing more sustainable alternatives. For some water authorities, LCA has become a key technique in the development of more sustainable supply, management and treatment systems.

LCA interacts with water systems in two distinct ways: 'water as impact' (water resource loss in producing goods or services) and 'water as a functional unit' (LCA of water service provision). First, water is recognised in many LCAs as a resource of such importance that it is used as an indicator of a system's environmental impact. In this context, the water indicator commonly represents all the uses of water over the life cycle of a particular system being analysed. In general, the water indicator is interpreted as an environmental impact that should be minimised. This interpretation can have its flaws, some of which are discussed below. Second, LCA can be used to assess the environmental impacts of water delivery and management systems (e.g. water treatment, stormwater management). LCA used in this manner aggregates a vector of environmental impacts over the life of a water system and then typically allocates impacts to a unit of consumption or disposal (e.g. global climate change impact per unit of water used or eutrophication impact per unit of wastewater treated). Incidentally, water use is often one of the environmental indicators considered (e.g. water used per unit of water delivered).

Although data collection and modelling may be complex and time-consuming, LCA has proved to be effective for assessing the sustainability of water systems and is increasingly used by water users and suppliers to maximise the sustainability of synthetic water systems. The following sections indicate elements of importance when using water as an indicator of environmental impact (see Section 8.2), and when undertaking an LCA of a water system (see Section 8.3). In Section 8.4, challenges and limitations in LCA for water assessment are discussed. Conclusions are drawn in Section 8.5.

8.2   Water as the indicator: a flawed measure?

As with any LCA, the goal and scope defines the objectives and desired outcomes of the study. Usually, the goal will be to determine potential environmental impacts associated with a particular system of interest. In determining impacts, the life cycle inventory is reviewed and indicators of environmental impact are calculated based on widely accepted scientific understanding. For example, the aggregation of carbon dioxide emissions may be undertaken as an indicator of the climate change effect of a system. In general, indicators typically used in LCA are measures of environmental damage or adverse affect, such as 'photochemical smog produced', or 'fossil fuels depleted'. Water as an indicator is used in a similar fashion to the latter – principally as a resource depletion measure.

Hence, the typical water indicator used in an LCA is constructed from the life cycle inventory by adding all water inputs to the system during its life cycle. The indicator provides an accurate measure of water inputs needed by the system, but not necessarily an indication of environmental damage. For instance, a system that accepts a given amount of water and discharges this water unchanged (such as a pipeline) will attract the same impact assessment in the water indicator as a process transforming a given amount of water into unusable sludge.

Where the inventory of water consumption is presented as an indicator of impact, this indicator also fails to deal with issues of local scarcity and site specificity. For example, the water impact of growing rice in southern New South Wales in Australia may be presented as similar to the water impact of growing rice in Bali in Indonesia, yet the local scarcity and implications of removal of this same quantity of water from different local systems may be quite different. It is therefore important to differentiate between inventory quantities (load, use or depletion) and effects (impacts of these quantities of use in a temporal and spatial context).

The potential for confusing an inventory quantity with actual environmental damage is not unique to water indicators. Other indicators meet with similar problems. For example, eutrophication describes nutrient disposal to the environment, with the potential to over stimulate biological growth in waterways. In some cases eutrophication will be limited by the phosphorous content of a waste stream, and in others it will be limited by the nitrogen content of a waste stream. The element that determines the eutrophication outcome depends on the geography of the locale. Nonetheless, eutrophication impacts typically do not incorporate spatiality, so the indicator may assume a worst possible eutrophication impact, which in reality may not occur.

Ultimately, water and eutrophication indicators assess potential rather than actual environmental damage. In the case of water, the range of actual impacts is possibly wider than for other indicators of more predictable environmental damage. For water, the assessment phase of the LCA process becomes crucial when drawing conclusions, especially when systems are being compared.

Alternatives to the 'coarse' measure of aggregate water consumption have been proposed in order to aid impact assessment. These alternatives characterise water inputs to a system more specifically. One such method characterises water inputs discretely into categories of use, consumption and depletion, where:

- 'use' indicates that water resource quantities are utilised and then made available to others
- 'consumption' indicates that the water resource quantities are denied to others
- 'depletion' indicates that water sources are either not renewed by the hydrological cycle or cannot be sufficiently replaced at the same rates that they are used by the natural hydrological cycle. (Owens 2002)

More detailed inventory data is required to support the characterisation proposed by Owens, but it does appear to provide some potential solutions. It is reasonable to acknowledge that water may be used multiple times before it is finally discharged to the environment, but this does not necessarily mean that the aggregated water indicator should be altered.

An alternative to Owens's approach is to retain the aggregate indicator of impact and to expand the system being analysed to include other processes using the water discharged by the system of interest. Water impacts would then be allocated to each system using the same body of water input. This method is potentially more involved, yet may also generate a level of rigour in the analysis that would ensure 'water made available to others' was usefully evaluated. This system expansion method is also consistent with ISO 14040.

Figure 8.1 illustrates one way to approach expansion of the system boundary. In this example, butter manufacture is the process of interest and, typically, the system boundary

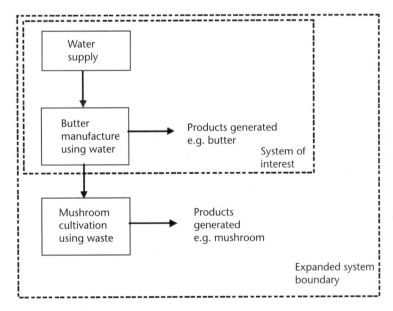

Figure 8.1   System boundary expansion.

would extend around this process alone. Assessment of water impacts would result in all the water used in the butter manufacture being allocated to a unit of butter production. In this example, wastewater is reused in a downstream process (because the butter process does not 'deplete' the water severely) and so the system boundary can be expanded to include the mushroom cultivation which is using the wastewater from the butter manufacture. The expanded system boundary allows the total water consumed to be allocated to both co-products (butter and mushrooms), and this more accurately reflects the actual consumption of the water resource.

Neither of these options deals with the location-specific issues of water use and nutrient flows. While no firm proposals have been put forward, differentiating water use within different archetypical climates would be an advance on the straight addition of water use. The impacts then of water extraction in water-limited environments could be assessed differently to water use in non water-limited environments. A measure could therefore consider water use in terms of the catchment area required to yield the water used. This would link water use to water availability; however, this method would also require some default assumptions about water use from larger regions or from unknown origins.

8.3   Assessing the life cycle impacts of synthetic water systems

Drought in large parts of populated Australia has led to scrutiny and introspection of water supply and consumption systems and their purposes. As water users and suppliers seek to plan within the physical limits of existing water resources, efforts have been made to seek alternative supply options and reduce demand. As a result, a myriad of water saving options have been developed, each theoretically capable of partly addressing water shortage, all other things remaining equal. With a wide range of options available, choosing the most appropriate options has proved difficult. Without selection criteria, it is difficult to decide which system to apply to the 'water problem'. LCA and Life Cycle Costing (LCC) have emerged as useful techniques to aid the choice.

In addressing water 'scarcity', an underlying theme is to make water supplies more 'sustainable'. Scarcity and sustainability are, of course, relative and related terms when used in this context. Solutions to water scarcity are typically presented as technically feasible initiatives selected purely on economic grounds, in the same way that many other capital investment projects are assessed, although many water practitioners, perhaps aware of the tight connections between water supply systems and the natural water cycle, have sought to incorporate sustainability into the decision-making framework. LCA has been chosen by some practitioners as a proxy for environmental sustainability, and in some cases this has led to a reassessment of the technical feasibility of the water option being considered.

The natural 'water system' is essentially cyclic and driven by gravity and solar power, yet there are numerous options for interfering with the habit of water molecules as they make their way around the various natural flows and stock points. Hence, the technology options for supplying and treating water vary and LCAs of these also vary. In essence, however, the analysis of a water supply or treatment system typically involves developing an inventory of impacts associated with the construction and disposal of the water system infrastructure, and an inventory of impacts associated with operation over its life. In combination, the infrastructure impact and the operational impact represent the total life cycle impact of the water system. Flow of water through the environment makes it difficult to understand completely the operational impacts. For this reason, many water LCAs attempt to capture extended consequential effects through expanded systems boundaries. It is therefore common for an LCA of a synthetic water system to include the downstream impacts on water treatment and stormwater systems, and in some cases supply systems.

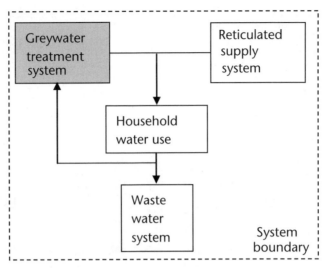

Figure 8.2 Systems that need to be considered when considering greywater system impact.

LCA typically commences with characterisation of a 'typical' existing system, consisting of potable supply and conventional treatment. However, supply varies according to terrain and scale, and treatment utilises a variety of physical, chemical and/or biological processes to remove the waste products from the wastewater stream. Up to four treatment stages are generally recognised. Primary treatment involves the removal of sand, grit, and other settable solids, oils, grease and fats from the water stream through the use of screening and sedimentation systems. Secondary treatment involves the removal of biological content such as human waste, food waste and detergents through the use of fixed film or suspended growth systems in the stream. Tertiary treatment involves the removal of residual toxins and nutrients such as phosphorus and nitrogen through the use of filtration, lagooning or constructed wetlands. If the treated wastewater is to be used where there is the risk of direct human contact, a disinfection stage is typically added involving chlorination, ultraviolet (UV) light or ozone (O_3) treatment. Hence, LCA characterisation of a 'base' system can be involved and is invariably unique to each case.

Alternative systems then require additional options to be considered either in addition to or in lieu of the existing/base system. The example of a household greywater treatment system illustrates a broadened system boundary. Upon initial consideration, the life cycle impacts of adding a greywater treatment system appear to be the production and operation of the greywater system itself. Unfortunately, limiting system analysis in this way does not address the implications of reducing reticulated water to the household and (potentially) reducing wastewater treatment requirements. To fully understand the impact of an alternative water supply system such as greywater treatment requires the LCA practitioner to think broadly about a group of interrelated system impacts, not just the system of interest (Fig. 8.2).

Once the system boundary is defined, flows of resources and resultant emissions can be characterised for both the system construction (i.e. the infrastructure establishment) and system operation. Using this approach, Yarra Valley Water has commissioned research investigating the life cycle impact and cost implications of different water servicing options. Figure 8.3 illustrates how the proportion of life cycle impacts associated with infrastructure varies for a series of water options considered for a particular reticulated suburban supply and disposal application ('global warming' indicator shown). However, there are specific and different issues in characterising assessment of infrastructure and operation, and these are discussed in the following sections.

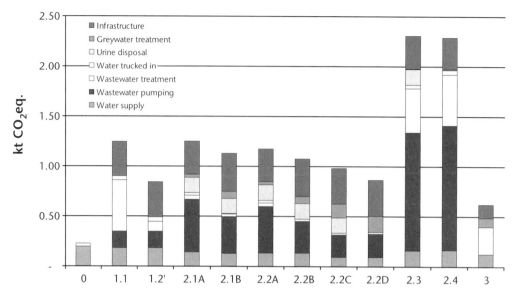

Figure 8.3   Infrastructure greenhouse impacts versus operational impacts for a series of supply and disposal options (Grant and Opray 2007).

8.3.1   Infrastructure

Infrastructure in water systems usually comprises tanks, pumps, pipes, foundations, excavation, catchment devices, settling ponds and other similar components. To determine the impact of such elements, detailed design information is required about these components and how the system will function. Understanding the detailed design of the water system to be analysed often takes as much effort as undertaking the LCA itself because most water system designs are not simple, especially when a large system boundary is drawn.

Two contrasting examples of water supply system designs are shown in Figure 8.4. They illustrate a limitation of LCA that is often overlooked. Typically, LCA is used to assess a design

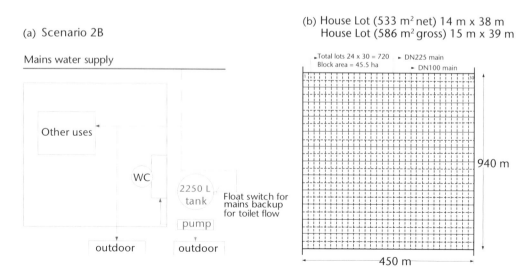

Figure 8.4   Design variations for water supply systems: (a) simple tank system design (Hallmann *et al.* 2003); (b) extract of suburban reticulated network design (Sharma *et al.* 2005).

Table 8.1   Main components and mass requirements of 600 and 2250 litre water tank assemblies (Hallmann *et al.* 2003).

	Material	600 L	2250 L
Tank body	Coloured LLDPE powder	35 kg	55 kg
Brass overflow protection	Copper (prim)	0.2 kg	0.2 kg
Brass outlet	Copper (prim)	0.35 kg	0.5 kg
Inlet strainer	LLDPE AU	0.35 kg	0.5 kg
Coloured LLDPE powder	LLDPE AU and pigment	35 kg 35 kg	55 kg 55 kg
Pigment	TiO$_2$	0.35 kg	0.55 kg

AU, Australian; LLDPE, linear low-density polyethylene; TiO$_2$, titanium oxide.

rather than to assist in creating one, so alternative supply or disposal systems need to have been designed before an LCA can be undertaken to compare them. Furthermore, if two contrasting designs are developed as in Figure 8.4, then implications flow for the definition of the system boundary and the functional unit, and for materials specifications.

The wide variety of materials and components necessitates an extensive range of background information. An inventory of common materials such as polyethylene, polyvinyl chloride, ductile iron, concrete and steel needs to be created in order to calculate the infrastructure impact of the system based on quantities identified in the design. Where good inventory data already exists for materials, the bulk of the research effort is associated with estimating design-specific quantities. For example, in Table 8.1 some component mass requirements for two domestic water tanks are illustrated.

Energy is invariably also required to run infrastructure. Pumps are often used in water systems and can be challenging items to inventory. Useful sources for materials used in pumping devices can be environmental declaration statements published by manufacturers. Figure 8.5 illustrates the major elements of a pump inventory where 'global warming' is the environmental indicator.

Installation and construction aspects of systems may also be important. Many LCAs do not include the impacts of energy and fuel used in the assembly of a mechanical plant used to produce infrastructure elements because they are typically minor compared to the impact of manufacturing the materials. Measurable impacts of water systems are often associated with energy and fuel for on-site excavation to embed pipes and tanks. Estimation of excavation impacts can be difficult and requires a detailed study of the process. A recent study of excavation practice associated with water infrastructure in Melbourne used data such as pipe diameters and trench widths as well as existing inventory data for transport and gravel extraction to estimate excavation impacts for pipework installation.

In water infrastructure models, it is also important to address system durability and lifespan, and likely disposal at end-of-life. Generally in a water LCA, a different operational 'design' life should be defined for each subsystem. An example of the lifetime assumed in a suburban water servicing model is shown in Table 8.2. Unfortunately, there is no generically applicable rule for the length of a system's life, which depends on the unique circumstances of the LCA being undertaken. It can be difficult to determine a new system's life, so a sensitivity analysis should be used to check the influence of 'life time' on the LCA results and conclusions.

Having defined the infrastructure aspects of the water system, it is possible to assess the operational impacts of the system, which for many LCAs will drive the total system impact.

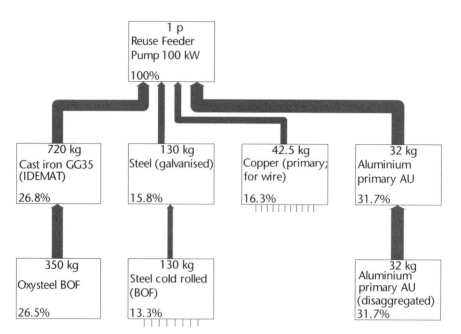

Figure 8.5   Global warming impact breakdown of a sewerage pump (Hallmann *et al.* 2003). IDEMAT, LCA database; AU, Australian Life Cycle Inventory dataset; BOF, Blast Oxygen Furnace.

8.3.2   Operation and consumption

Operational impacts of a water system predominantly relate to demand and are therefore seasonal. Notwithstanding seasonality, the impacts of water servicing systems are typically driven by water transport energy requirements (usually pumping) and water treatment requirements. Assessment of water treatment impacts usually requires assessment of energy requirements, chemical additives and chemical and biological emissions to the environment.

Pumping in water systems is typically undertaken with electric motors and centrifugal pumps, although other methods exist. Pumping energy (P) can be calculated in several ways, including:

$$P = \frac{\rho g Q h}{1000 \eta}$$
Equation 8.1

where: 'ρ' refers to the density of water which is 1000 kg/m^3, 'g' refers to the gravitational constant which is 9.8 m/s^2, 'Q' refers to the discharge which will be 1 m^3/s, 'h' refers to the pumping head, which also incorporates the efficiency loss of the pipe, and 'η' refers to the combined pump and motor efficiency.

Table 8.2   Assumed life times for subsystems in a suburban water-serving model (Grant and Opray 2005)

Infrastructure	Life time modelled (years)
Household plumbing and infrastructure	25
Trunk mains (water and sewer)	70
Non-truck water and sewer mains	100
Water treatment facilities	60

Equation 8.1 illustrates that the power used by pumps in water systems is based on flow rate and pumping head.

Pumping energy requirements can also be determined using 'energy maps'. Energy maps describe the energy requirements of reticulated water systems; both water supply and sewerage. Energy usage tends to be related to topography, with more mountainous terrain requiring more energy for water services (supply and sewerage). An energy map for sewerage for a sample of geographic regions in Melbourne is shown in Figure 8.6 and for water supply in Figure 8.7.

When undertaking an LCA for a reticulated system, it is often useful to calculate operational energy both via a 'first principles' approach (a power formula as shown in Equation 8.1) and a comparison of the results with those obtained by using an energy map. By investigating any inconsistency between the two results, errors can be identified and rectified, improving confidence in the overall result. Operational environmental impacts often relate significantly to energy requirements of pumping, consumption of operating material, and the contaminant load of the effluent stream (emissions to the environment). The operating material consumption of a treatment system consists of those materials, such as chlorine, filter sand, flocculating agents, pH adjusters and so on, that are consumed in the operation of the treatment system. In an LCA of a treatment system, the manufacturing and delivery impacts of these chemicals need to be included in the life cycle inventory. In many cases these 'consumed' materials often become emissions to the environment, albeit in transformed states. Noting that various treatment systems perform differently, some examples of effluent stream contaminant balances used in a recent LCA study are shown in Table 8.3.

Unfortunately, the effluent quality of a treatment system is not solely determined by the treatment system itself, but also by the quality of the influent stream. For this reason, it is

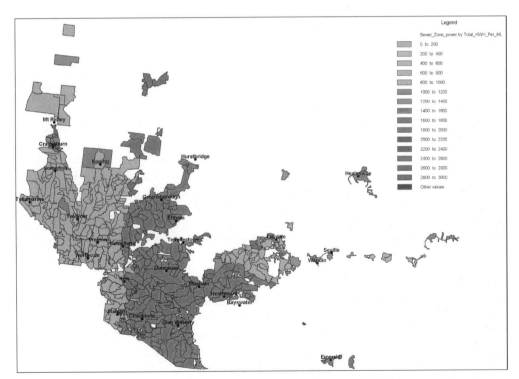

Figure 8.6   Energy map for sewers (kWh per ML) (Yarra Valley Water 2005).

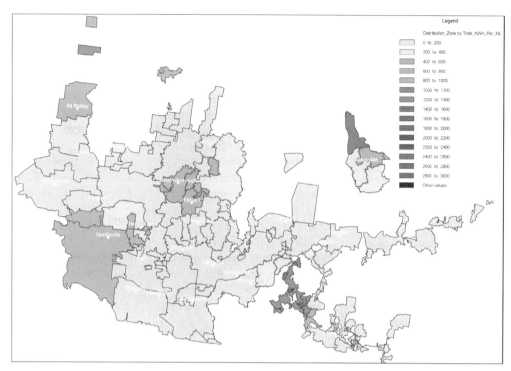

Figure 8.7   Energy map for water supply (kWh per ML) (Yarra Valley Water 2005).

usually necessary to undertake some analysis of water and contaminant balance to ensure that influent loads are appropriate to the treatment systems being considered, and that contaminants are assessed throughout the water system. Having undertaken a contaminant balance and water balance and confirmed that the treatment system of interest will likely deliver the effluent quality required, it is still important to consider potential changes in the influent quality.

LCA models are often constructed using fixed 'per litre' effluent qualities to determine treatment system emissions, such as the fixed quantities indicated in Table 8.3. These can lead to errors if interventions alter effluent qualities. One example would be the addition of a urine separation system to a household wastewater stream in order to reduce nitrogen loads in the wastewater. Clearly, in this case, an adjustment in the assumed effluent would be required in order to model impact reductions accurately. In addition, apart from pumping energy and operational material consumption, the major impacts of water treatment systems typically occur under the eutrophication indicator, which shares many of the weaknesses of the water indicator.

8.4   Discussion: water systems, design and social context

Clearly, to determine the impacts associated with infrastructure and operation requires a distinctly different approach in each case. Infrastructure impacts tend to be driven by the quantity and type of material used in construction, whereas operational impacts tend to be associated with energy consumption of pumps and other transport devices. Although materials are consumed in operation, energy (electricity derived from liquid fuel) tends to dominate during this phase. There are also other issues associated with water systems that affect overall envi-

Table 8.3   Contaminants in effluent streams of various treatments systems (assuming household wastewater influent) (Grant and Opray 2007)

System	BOD (mg/L)	SS (mg/L)	Ammonia nitrogen (mg/L)	Total nitrogen (mg/L)	Total P (mg/L)
Reticulating Biological Contactor – Effluent	10 to 20	15 to 20	2	10 to 20	0.3
Recirculating Trickling Filter	<10 (field reports of <5)	<10 (field reports of <5)		<25 (20 used)	
Living Machine	≤10	≤10	≤1	<10	<5
Re-circulating Sand Filter	≤10	≤10		≤50%	
Aerobic Lagoon	20 to 40	40 to 60		Varies with climate – 10 to 20% removal	Varies with climate – 10 to 20% removal
Faculative Lagoon	30 to 40	40 to 100		Varies with climate – 1 to 60% removal	Varies with climate – 1 to 50% removal

BOD, biological oxygen demand; P, phosphorus; SS, suspended solids.

ronmental performance. Here, discussion focuses on the dynamics and design of water systems, and the sociotechnical linkages they exhibit.

Water supply system design is rarely based on an average usage period. Water systems, by their nature, will be sized according to peak demands for water over a period rather than average demand. For this reason, quoted average usage rates are generally of little value when determining the infrastructure impacts of a water system. An example is household water demand in Australia, which is in significant part driven by garden watering requirements. The quantity of water applied to a garden can be as high as 50% of total household water consumption, although it is roughly 35% in the average Melbourne home, or about 70 kilolitres to 75 kilolitres per year. Due to lack of data, it is difficult to determine more meaningful figures to account for varying garden sizes, relative sizes of garden features (e.g. lawn and concrete), plant types and social practice.

Table 8.4 illustrates the large swing in demand for water driven, for example, by garden requirements. If a water system is to supply this demand, it will need to be able to provide maximum capacity in the summer and the same capacity will not be required in winter. Seasonality in operation therefore needs to be modelled at a relatively short time interval during operation in order to capture system supply variation when multiple systems are in use. In a mixed-mode greywater and mains supply system for households, during different seasons, water will be exchanged from one system to the other and vice versa. The operational share of supply must be understood in order to calculate impacts.

The challenge for some water supply systems is that capacity to supply water is lower when demand is higher, due to the seasons, as in the case of rainwater collection systems (see Fig. 8.8), or is constrained below the peak requirement, as in the case of a greywater treatment system. The rainwater collection example illustrates the need to undertake system analysis at a relatively fine time interval (usually daily) to ensure supply and demand mismatches are properly comprehended.

Table 8.4   Seasonal water usage data (Hallmann *et al.* 2003)

Season	Share of annual garden use	Average amount of water applied/garden/day (L)
Summer	66%	536.7 (90 days)
Autumn	22%	177.4 (92 days)
Winter	0%	0 (92 days)
Spring	12%	93.3 (91 days)
Total	100%	Total annual water use: 73.1kL

Mismatches in water systems are usually dealt with by adding storage capacity, although 'behavioural demand side solutions' are becoming more common. Analysis of daily supply and demand for water allows impacts of various supplementary tank sizes to be considered. Usually, environmental impacts are minimised in supplementary water systems when only a portion of water is supplied through the supplementary system. This luxury does not exist when a water supply system is isolated from reticulated supply and must carry the expected demand throughout the year.

The influence of social practice extends beyond user 'behaviour', since the size of the water supply system determines capacity for future use – and there is a tendency for capacity to be fully used. New uses may also be found for extra capacity. Hence, where new supply sources such as desalinisation plants are envisaged, with significant, long-term investment in new infrastructure and supply capacity, new consumption outlets may be identified. Following the drought period which provides the impetus for such infrastructure investment, new uses for the water and infrastructure are found, in the same way that car ownership leads to more vehicle kilometres travelled.

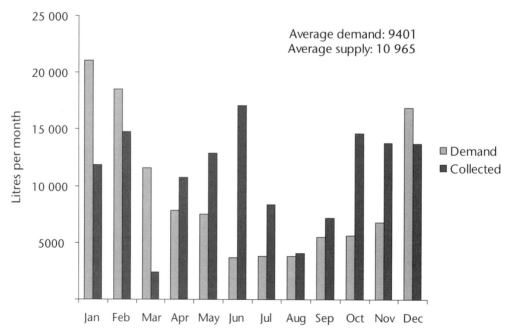

Figure 8.8   Demand for water (toilet and garden only) versus total water collected on a typical suburban roof (calculated by the authors from data in Hallmann *et al.* 2003).

This 'ramping up' of demand through complex infrastructure-utilisation relationships has been extensively investigated in sociotechnical literature, where it has been shown that stakeholders prescribe appropriate levels of demand into systems of provision, thus shaping consumer behaviour (e.g. Chappells and Shove 2004). Oversizing water infrastructure systems can therefore be expected to lead to increased use of latent capacity. Hence, seasonality and appropriate sizing are important in a water-constrained environment. LCA as currently deployed has little to say about these interrelations between design and environmental impact outcomes, since the implication of sizing is generally considered to be simple – more impact through more materials and energy used for the oversized system. Moreover, the general assumption is that each unit of consumption is equally valid; indeed, the functional unit is invariably 'per unit of water consumed', which does not account for total consumption or types of consumption. There is ample scope to extend LCA in these areas.

8.5   Conclusions

As society begins to recognise the finite nature of water resources, LCA emerges as a tool to assist development of more sustainable water practices by water supply authorities and water users. Water is handled by LCA in two distinct ways: first, to characterise the potential environmental impact of any system being analysed; and second, in the analysis of impacts of water supply, management and treatment systems.

Water as an indicator of system impact must be interpreted carefully to ensure that real environmental impacts are highlighted. Aggregate water use is an inventory item rather than a direct representation of environmental damage, although the process of LCA, as defined by ISO 14040, states clearly that impact indicators must only be used with interpretation. The potential spectrum of actual environmental impacts can be carefully addressed by a thorough and logically presented interpretation of water indicator results. Although alternative assessment methods are available to reduce the need for interpretation, in many cases they can be avoided by expanding system boundaries to capture complicating systems such as other users of the same water resource.

The use of LCA to analyse water systems requires a broad perspective and a wide system boundary. The multitude of interactions between water servicing systems requires both supply and treatment systems to be considered when assessing a water saving or enhancing design. Determining the life cycle impacts of water systems requires a detailed understanding of system design, both its infrastructure and operation. Infrastructure designs need to be well documented in order to capture the many materials and installation activities of the system being considered. Detail is also required to understand likely operational impacts, including energy use, material consumption and emissions.

In many cases, system infrastructure and operational behaviour are influenced by the seasonal nature of water demand. Understanding seasonality often necessitates detailed modelling of water demand and supply capacity. Modelling usually has to be undertaken at a fine time interval to ensure that any shortfalls in system supply capacity are fully comprehended. As indicated above, seasonal water demand fluctuations can have a direct impact on system sizing, driving larger capacity systems than average data would suggest is necessary. Furthermore, water supply systems provide a good example of the interdependence between system design and use. Where 'unlimited' use is enabled at the turn of a tap or the creation of well-positioned drains, then what is considered to be a 'normal' rate of consumption is likely to rise, with boundaries of normality being redrawn irrespective of previous 'behaviour' conventions. 'Drought' may then lead to these boundaries being (abruptly) redrawn. Thus, drought is at once both a function and regulator of demand and supply systems, and incorporation of such factors into design, sizing and optimisation are fertile future territories for decision-making for water service provision.

Although LCA can be complex and time consuming, it provides an objective and often surprising environmental perspective. In requiring a detailed approach, it provides a cross-check of the water system design that can reveal design shortcomings and potential efficiency improvements. More importantly, however, it provides an objective sustainability assessment, that if acted upon at the design stage will ensure that the water systems of the future operate successfully and sustainably in the long term.

8.6   References

Chappells H and Shove E (2004) Infrastructures, crises and the orchestration of demand. In: *Sustainable Consumption: The Implications of Changing Infrastructures of Provision.* (Eds D Southerton, B Van Vliet and H Chappells) pp. 130–143. Edward Elgar, Cheltenham, UK.

Grant T and Opray L (2005) 'LCA report for sustainability of alternative water and sewerage servicing options'. Yarra Valley Water. RMIT University, Melbourne.

Grant T and Opray L (2007) 'Sustainability of alternative water and sewerage servicing options – life cycle assessment stage 2'. Centre for Design, RMIT University, Melbourne.

Hallmann M, Grant T and Alsop N (2003) *Life Cycle Assessment and Life Cycle Costing of Water Tanks as a Supplement to Mains Water Supply Melbourne.* Centre for Design, RMIT University, Melbourne.

Hardin G (1968) The tragedy of the commons. *Science* **162**, 1243–1248.

Houston C (2006, November 7) 'Phillip Island's drinking water may soon run out.' *The Age.* Retrieved 18 March 2008 from <http://www.theage.com.au/news/national/phillip-islands-drinking-water-may-soon-run-out/2006/11/06/1162661617861.html>.

Owens JW (2002) Water resources in life cycle impact assessment. *Journal of Industrial Ecology* **5**(4), 37–54.

Sharma A, Grant A, Tjandraatmadja G and Gray S (2005) Sustainability of Alternative Water and Sewerage Servicing Options. CSIRO Urban Water, and Centre for Design, RMIT University, Melbourne.

Smith A (1776) *The Wealth of Nations.* JM Dent and Sons, London.

Smith DI (1998) *Water in Australia: Resources and Management.* Oxford University Press, Melbourne.

Yarra Valley Water (2005) *Water Energy Map.* Melbourne.

Chapter 9

Life cycle assessment and agriculture: challenges and prospects

Ralph E Horne and Tim Grant

9.1 Introduction

Food is a basic human need, and the production of food is a considerable activity. The vast majority of our food involves farming – a practice that has endured for thousands of years. This longevity may suggest that agriculture is an essentially 'sustainable' activity. The inevitable wealth of experience associated with such a well-established industry suggests there has been ample opportunity to test and develop sustainable practices. However, there are several reasons to question the sustainability of different agricultural practices. Food and drink production – one of the outputs of farming – is a significant contributor to environmental impact. For example, in Australia, food represents 35% of the ecological footprint (Global Footprint Network and the University of Sydney 2005).

Technologies for and the scale of agricultural endeavour have changed radically over recent decades, with resultant widespread land use conversion and increasing demand for a range of fossil-fuel based agricultural inputs. Water and land resources are often recognised as local or regional impact issues, and the global contribution to greenhouse gas emissions from farming is significant, at some 8.5–16.5 Pg CO_2-eq y^{-1} or 17% to 32% of the total, with major contributions to this coming from nitrous oxide (N_2O) emissions from soils and enteric methane (CH_4) emissions from livestock (Bellarby *et al.* 2008). However, the largest single contribution comes from the clearing of native vegetation for agriculture rather than directly from tillage. Cropland soils have relatively low carbon concentrations, and this raises the prospect of agriculture shifting from being the second largest sectoral emitter of greenhouse gases globally to being a major carbon sink in the future, if framing practices can be changed to reduce fossil fuel use and build carbon concentrations in soils. Such an outcome will require considerable changes in practice since, for example, global CH_4 and N_2O emissions increased by 17% from 1990 to 2005, and N_2O emissions are predicted to rise by 50% over the period 1990 to 2020, given current trends (US EPA 2006).

World food prices are rising, which may indicate that the capacity of supply to meet rising demand is dropping. Climate change is also contributing to the failing sustainability of agriculture through water shortages, fiercer and more intense droughts and soil loss, compounded by increased competition from crops grown for biofuels and rising meat consumption. As a major producer (and exporter) of agricultural products, Australia has a strong case for focusing on the environmental impacts of agriculture. State of the Environment (SoE) reports indicate that agriculture is a major consumer of rural resources in Australia (SoE 2006). As indicated in

Chapter 5, agriculture accounts for 62% of land use and some two-thirds of Australia's water use. So, how may life cycle assessment (LCA) assist in assessing impacts and driving the uptake of more sustainable techniques in agriculture?

In this Chapter, case studies are used to illustrate some of the key developments and applications of LCA to agriculture in Australia. The initial challenges for application of LCA to agriculture are introduced in Section 9.2. In Section 9.3, sugar cane production in Queensland is used to explore optimal land uses and functional aspects of agricultural production within a socioeconomic context. In Section 9.4, an example from dairy farming is used to illustrate the importance of taking a 'whole-of-system' view of water use in milk production and consumption. In Section 9.5, the balance between agricultural production and food processing is again explored, this time through an example of maize production and corn products. In Section 9.6, examples from 'food miles' and other 'farm-fresh produce' debates are reviewed in the context of LCA in agriculture. In Section 9.7, discussion focuses on the methodological and research challenges in application of LCA for policy and decision-making in agriculture, while Section 9.8 concludes with a brief consideration of the future potential utilisation of LCA in the context of farming in an environmentally constrained world.

9.2   Issues in the application of LCA to agricultural systems

There is generally a lower level of maturity in the methods and techniques of LCA regarding the assessment of agricultural systems compared to industrial systems. The setting of system boundaries is often more complex, as agricultural systems are relatively open, involving land, biodiversity, and a range of interrelated hydro-bio-chemical processes. The development of reliable datasets in LCA generally relate back to its origins in energy assessment and analysis of manufacturing processes. Hence, data relating to biodiversity change, soil-crop dynamics and other aspects of agricultural systems are generally less available in LCA inventories than energy data required for industrial processes.

Although agricultural LCA data is becoming more widely available, data shortages continue to challenge the widespread use of LCA for agricultural practices. Moreover, the diversity of these practices and the variables involved are significant. Even for a single agricultural product, there is typically a range of different types or modes of production. For example, meat production may involve different species farmed at different intensities using varying amounts of land and water, with different patterns of inside/outside husbandry, in widely differing climates and with widely differing technologies, types and amounts of inputs (food, fertiliser, infrastructure and fuel) and outputs in the form of products and by-products (e.g. prime cuts, other cuts, mechanically recovered fractions, tallow, bone products, hide products and offal). Of course different system characteristics and components would apply in the case of game meat. Similarly, there are wide ranges in variables for different dairy systems and crop systems.

Even where a single system type can be identified and data secured, significant variations in actual LCA results may be achieved within conceivably similar parameters, across different situations. Industrial systems for producing a particular product from a particular material may typically involve a narrowly predictable range of operating conditions within a well-defined (indoor) system. However, in agriculture, within a single field, soil moisture variations may determine widely different soil greenhouse emissions, or different tenants on neighbouring farms may adopt varying practices in fertiliser or pesticide application from year to year. In other words, LCA is most readily applied to predictable, replicable, closed systems, which agriculture is not.

Nevertheless, LCA has already made contributions to the environmental assessment of agricultural systems. Notably, it has been used to identify so-called 'counter-intuitive' results, indicating that 'natural' (crop-derived) products often cause more environmental burdens than 'synthetic' manufactured alternatives. Moreover, agriculture generally is becoming more industrialised and systematic and, while this is not necessarily environmentally benign, the increasing use of digital technologies and automated factory-style processes on-farm holds out the possibility of better data and potentially improved LCA accuracy. Concurrently, there is rising interest in the environmental (and especially greenhouse and water) impacts of food and agriculture, partly in recognition that food is a major component of the eco-footprint of most households in the developed world. Since food is a basic commodity that invariably and increasingly involves agricultural technologies and practices, the systematic approach of LCA to assessment of the various environmental burdens across different agricultural options is a logical response to questions about the relative environmental impact of different foods.

9.3   Sugar cane: options and optimisation

In the work of Marguerite Renouf *et al.* (e.g. Renouf 2006; Renouf 2007; Renouf and Wegener 2007; Renouf *et al.* 2005), sugar cane production systems in Queensland are analysed and compared to other sugar production systems, and the use of sugar cane crops as a substrate for other products is also considered through LCA.

One of the key challenges of scoping this work was the difficulty in defining a functional unit. First, from a farm perspective, the functional value of the crop is return on investment from a given area of productive land. While most capitalist enterprises have this as a primary function, what makes agriculture different is the flexibility of farm capital (e.g. land, tractions and water) to switch between crop types and rotation systems. Second, from the sugar perspective, sugar cane has multiple production alternatives for both the sugar syrups and the residual cane fibre. The issue then becomes an assessment of the optimal use of the land, given the alternative ways of producing each of the product options available from cane production.

The initial goal and scope of the work included characterisation of the system of sugar production from sugar cane based on production practices in Queensland (which accounts for about 94% of Australia's total production). The scope used is 'cradle-to-gate' and the functional unit is a tonne of raw sugar leaving the mill gate. Life cycle impact assessment results were generated for the following impact categories, using the impact assessment model Eco-Indicator 95 (Goedkoop 1995, cited in Renouf 2006):

- energy input, as megajoules of fossil-fuel energy (MJ)
- greenhouse gas emissions, as kilograms of carbon dioxide equivalent (kg CO_2 eq)
- acidification potential, as grams of sulphate equivalent (g SO_4^{-2} eq)
- eutrophication potential, as grams of phosphate equivalent (g PO_4^{-3} eq)
- water use, as kilolitres of fresh water (kL).

Initial findings indicated that crop production generally dominates as a source of environmental burdens, and two problems are identified here. First, there is apparent variability in crop production systems within Queensland, and so three scenarios were chosen to explore and reflect this variability, as follows:

1. a 'state average' farming system consisting of area-weighted state averages for cane yields, inputs, harvest practices and transport
2. a wet tropics scenario reflecting cane growing in north Queensland, with relatively low nitrogen input, nil irrigation, and lower cane and sugar yield
3. a high yield scenario reflecting cane growing in the Burdekin region with relatively high nitrogen input and irrigation.

This three-scenario approach, with an estimated average and two 'extremes', provides a useful indication of variability.

The second problem is that many of the dominant environmental impacts from cropping systems are associated with dynamic processes in agricultural soils. This is a challenge for LCA due to the general lack of LCA inventory data relating to soil dynamics, and it indicates the importance of integrating agricultural soil modelling data with LCA. Specifically, Renouf (2006) derives emission rates for nitrogen through modelling, using the agricultural simulation model APSIM-Sugar cane, a specific sugar cane model designed to be able to be used to estimate nitrogen losses through soil denitrification and leaching under various soil and climatic conditions and for different trash management regimes. For this study, nitrogen (N) denitrification and leaching rates were derived from work undertaken by Thorburn et al. (2004) with further information from Brentrup et al. (2000), and this is supplemented with other studies (including Denmead et al. 2005; IPCC 1997; AGO 2003; Thorburn et al. 2005) to derive partitioning of soil nitrogen losses between nitrous oxide (N_2O), nitrogen gas (N_2) and nitrogen oxides (NO_x). Nitrogen emissions are expected to vary across the three scenarios, according to climatic conditions, soil type and agricultural practice (particularly trash blanketing, as practiced in the Wet Tropics). Moreover, all N_2O emission rates derived for this study are considerably higher than the generic 1.25% figure for arable cropping provided by the Intergovernmental Panel on Climate Change (IPCC), in keeping with other work on irrigated maize (Beer et al. 2005).

Allocation of inventory flows is another common issue for agriculture-related LCA, where co-products and/or by-products are produced from the same crop, husbandry or other agricultural practice. The co-production of raw sugar and molasses was taken into account using both economic allocation and system expansion. Using economic allocation, the impacts were allocated 96% to sugar and 4% to molasses based on production rates of 143 kg raw sugar and 26 kg molasses per tonne of cane, and average market values of A$300/tonne for raw sugar and $70/tonne for molasses. Using system expansion, the difficulty lies in the equivalency of the substitute co-products or by-products, and the efficacy of the data and subsequent values used in the analysis. In this study, substitution of molasses was assumed to be 40% by barley (pasture supplement), 20% by wheat (for fermentation for ethanol production) and 40% by nothing – as a significant proportion of molasses is used as an attractant in feed and is considered non-essential.

The results were almost identical for each allocation approach, suggesting that choice of allocation method is not an important issue for this system. Agricultural activities dominate the energy profile, in particular fertiliser production, on-farm fuel use for tractors and harvesters, and electricity for pumping irrigation water (where applicable). The milling of cane to produce raw sugar requires very little input of fossil fuel energy since the steam and power used in the mills is generated from bagasse combustion. The combined contribution of on-farm and cane railway capital goods was significant (between 5% and 10% of total energy input), and in regions where road transport is also necessary (as in the wet tropics), this aspect becomes significant too. However, the difference in energy input between the scenarios is not great considering the uncertainties. Cane production in the Burdekin has higher inputs per hectare (fuels, fertilisers and water), but also higher cane and sugar yields. Conversely, the wet tropics has lower inputs per hectare, but lower cane and sugar yields. Hence, on the important performance measure of ratio of inputs to yield, the energy results are of a similar magnitude.

Regarding eutrophication potential, emissions of nutrients to air (ammonia NH_3; nitrogen oxides, NO_x) and water (nitrate, NO_3^-, phosphate, PO_4) from sugar cane fields provide the most significant contribution, with differences in the results for each scenario attributed to different environmental conditions (climate, soil type), which influence field emissions; wetter areas (Wet Tropics) tend to have higher losses of N via denitrification, leaching, and ammonia volatilisation. Trash blanketing (as practiced in the Wet Tropics and state-average scenarios) also increases nitrogen losses, by providing additional nitrogen and a carbon source, both of which promote microbial and metabolic processes in the soil. The Burdekin region has the lowest result since it is a drier environment and cane is grown without a trash blanket. Emissions from sugar cane fields (ammonia and nitrogen oxides) generally make the most important contribution to acidification potential with environmental conditions again providing an important influence on the variation observed between different regions. Regarding water impacts, sugar cane responds particularly well to high water availability and some 60% of the Queensland crop is irrigated. This is an important environmental consideration in countries with low rainfall and poor supplies of irrigation water.

Importantly, the sugar cane study highlights areas where data improvements can be made, and notes three main factors found to influence variability in sugar cane environmental performance:

1. environmental conditions – climate, soil type, and topography
2. agronomic practices
3. geographic location relative to supporting infrastructure.

Although greenhouse gas emissions are influenced by both the intensity of resource inputs and environmental factors, there is considerable uncertainty around actual N_2O emissions between scenarios, and that actual variability may be greater than that reported by the limited data set used. Indeed, the results show the importance of field emissions in sugar cane production, particularly emissions of all nitrogen species (N_2O, NO_x, NH_3, NO_3^-). Eutrophication and acidification potentials appear to be influenced by environmental factors, but the retention of crop residues after harvest (trash) is also important. Regarding water impacts, not surprisingly, the key factor is irrigation.

The essential conclusion of the work by Renouf (2006) is that variability should be considered carefully in LCAs of agricultural crop production, particularly in relation to environmental conditions, but also agronomic practices. An agricultural practice 'mean' average performance assessment may therefore be less meaningful than an equivalent measure for a product manufacturing plant-based system where the variables are fewer and the controls over the system greater. This is a major challenge for 'traditional' LCA, which typically seeks to identify and model some form of 'standard' performance.

Two reflections on Renouf's sugar cane studies reveal further implications for LCA practice and potential. First, where there is variability across the system, as in sugar cane growing in Queensland, a series of LCA studies may be used to identify empirically a set of key variables that exist in practice and have the potential to significantly alter results. Examples may be soil moisture content, particular field practices and irrigation rates. From this, a tool may be envisaged where the implications of varying these factors can be quickly and with sufficient accuracy modelled by a non-LCA specialist. This provides the potential to capture local variations in practice and conditions in the modelling of performance, and where it is possible to vary inputs, practice may be altered according to desired performance outcomes. This approach holds out the possibility of bringing LCA forward temporally to become a field-scale tool that can alter design and practice of activities in real time (for further discussion on 'quick' LCA tools, see Chapter 11).

Second, as the essential issue is optimisation of the sugar cane industry, it is still open as to whether the resources involved – land, water, labour and various inputs – are optimised in sugar cane growing, or whether alternative practices are preferable. Often, LCA is conducted in a propositional or planning setting, whereas here there is a sugar cane production industry that has been well-established for over 100 years. Any change in sugar cane agricultural practice therefore has cultural, social and economic consequences for existing agricultural and related communities. An appropriate question might be: 'we have land, water allocations and farmers – what is the best thing to do?' But LCA can realistically only contribute to environmental dimensions of possible 'answers' to this question, in providing options for optimising the use of these available 'resources'. It is, however, a potentially powerful technique for application beyond the practice-driven 'optimisation' question to the policy-driven questions of what to do with particular limited resources such as land and water (and potentially, greenhouse gas emissions, under a carbon trading regime).

9.4   Tail wagging the cow? LCA of milk production and packaging

The dairy industry is significant in Australia, and the need to understand the life cycle impacts through various stages and products of the industry led to a study sponsored by Dairy Australia, with links to the International Dairy Federation and the United Nations Environment Programme (UNEP) (Nicol 2005). The research was undertaken by Sven Lundie and Andrew Feitz at the Centre for Water and Waste Technology (University of NSW) and Michael Jones at the Centre for Food Technology (Queensland Department of Primary Industries). The scope of the study is extensive, covering both a range of environmental impacts and the full life cycle, including by-products, as well as post-farm processing, packaging and transport.

Particular focus was placed around water, energy and greenhouse impacts, particularly on the split between on-farm and post-farm activities. The study was conceived partly out of concern that the packaging of milk is a significant environmental problem. However, correctly, the LCA was scoped to include these impacts across the life cycle, rather than assuming outcomes and focusing entirely on packaging.

The LCA results indicate that any pre-conceptions about life cycle impacts of dairy packaging were best treated with caution. For example, on-farm water use was found to be far more significant than that associated with packaging and manufacturing, with 99% of life cycle water use being consumed on the farm, and only 1% in manufacturing and packaging (Nicol 2005).

Notably, differences were identified between chilled/fresh and aseptic/ambient packaging options, in significant part as a result of the structure and configuration of the packaging and processing industry surrounding these technologies. The few aseptic/ambient packaging plants means that large transport distances are necessary and this leads to a relative rise in the environmental burden attributable to the system. Hence, in the 'market milk' (fresh milk) option analysed, energy use associated with packaging contributed 14% of the total, with 21% attributable to the farm, 14% to manufacturing, and 3% to retail transport. In the case of ultra high temperature (UHT) milk, these proportions changed to 19% for packaging, 13% to the farm, 18% to manufacturing, and 19% for retail transport. This particular infrastructure arrangement leads to a higher overall energy use for the UHT scenario compared to fresh milk, per tonne of product. As expected, the energy footprint of yoghurt and cheese is significantly higher per tonne of product than either milk scenario.

Unlike many LCA studies, the results for greenhouse gas emissions do not follow the pattern identified for energy use. This is principally due to the methane emissions associated with ruminants. As a result, in this study, some 52% of greenhouse gas emissions were found to be attributable to the farm, compared to only 13% to 21% of energy use.

This reported study is only one of many conducted on dairy production worldwide, and there is a high level of interest in the environmental impacts of dairy farming and products, reflected in the level of LCA activity, including subsequent studies led by Dairy Australia (e.g. Feitz *et al.* 2007). The range of emissions and activities involved in the complete dairy production chain presents complex challenges for the design of policy and practices to reduce environmental impact. Some impacts are more controllable than others and there is significant risk of burden shifting and/or inefficient outcomes where particular environmental issues in parts of the chain are targeted. For example, a support mechanism for UHT milk designed to reduce wastage may lead to higher transport emissions unless it were accompanied by reconfiguration of the UHT production system. Other policy implications are discussed further in Section 9.7.

9.5   Following maize from seed to chip: impact versus value?

The sugar cane and dairy case studies (Sections 9.3 and 9.4) illustrate some of the data and methodological issues raised by the application of LCA to farm-scale agricultural impact assessment. In this case study, we build further on the approach of looking beyond the farm gate, and examine the system of corn chip production from seed to retail shelf, including growing, transport, processing, manufacturing, packaging and distribution. This account is drawn substantially from Grant and Beer (2008), based on research led by CSIRO into the life cycle impacts of irrigated maize production (Beer *et al.* 2005; Edis *et al.* 2008; Kirkby *et al.* 2006; Meyer *et al.* 2006; Grant and Beer 2008). The Grains Research and Development Corporation and the Australian Greenhouse Office contracted CSIRO, the CRC for Greenhouse Accounting and the University of Melbourne to undertake the study.

Although the impacts assessed are limited, with a particular focus on life cycle greenhouse gas emissions, the study takes a system that is more extensive than many agricultural LCAs. First, it incorporates primary research on soil greenhouse gas balances, including emissions associated with additions and changes in soil nitrogen. Second, it extends the system to include processing of the agricultural product into a final retail food, and includes packaging of the food. Third, it includes an investigation of economic value (and specifically, the value added during each production step). In keeping with the extended system boundary, one 400 g packet of corn chips is adopted as the functional unit, with the measurement unit being kg CO_2 eq per packet of corn chips (see Fig. 9.1).

Irrigated summer crop maize was chosen as it has been identified as a potentially strong emitter of greenhouse gases, particularly of nitrous oxide (N_2O), since fertiliser use leads to N_2O emissions (Flessa *et al.* 2002), and the maize industry uses high rates of fertiliser application. Data was identified as a key issue early in the project, and the project team was in Griffith in February 2004 to inspect the farm, and then to obtain relevant energy-use data by visiting the Bendigo factory of the corn chip manufacturer. Recognising that farms may use different irrigation techniques, different soil management techniques, and different fertiliser application regimes to those used on the field site, consultation was undertaken with growers in the field in order to determine the representativeness of the study parameters being examined on the experimental farm (to inform the uncertainties associated with the life cycle estimates).

On-farm measurements of N_2O emissions from nitrogen fertiliser applied to maize crops were conducted at Commins Brothers property at Whitton, New South Wales (34.5°S 146.2°E). The measurements were conducted on a site already established for 5 years, to investigate the interactions between nitrogen and stubble retention on soil carbon dynamics. The fluxes of N_2O and CO_2 were measured on three of the established treatments:

- zero N fertiliser application and stubble removed by burning
- 329 kg N/ha, stubble removed by burning
- 329 kg N/ha, stubble mulched and incorporated into the soil.

The basic LCA was for irrigated maize supplied to corn chip producers. It was assumed that:

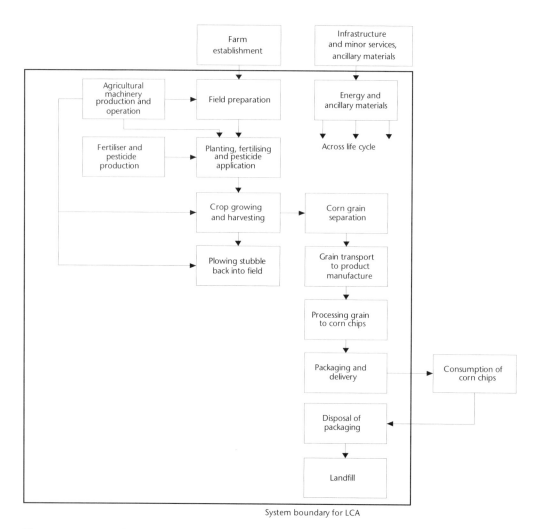

Figure 9.1   System boundary (boxed) for life cycle assessment for corn chip production (after Grant and Beer 2008).

1. about 50% of the crop was produced by conventional cultivation (i.e. with stubble burned) and 50% with stubble incorporated
2. the irrigation water was supplied by gravity feed from irrigation channels (35%) and from bores (65%) with an average depth of 50 metres.

With these assumptions, it was found that for the corn chip production chain, the total net emissions per 400 gram packet of corn chips reaching the domestic market are 0.53 kg CO_2 eq. This comprises 68% CO_2, 30% N_2O and 4% CH_4, with a 2% greenhouse gas credit for carbon sequestration of the cardboard packaging in landfill. The single largest source of greenhouse emissions is the emission of N_2O on the farm as a result of fertiliser application (0.126 kg CO_2 eq per packet). The next largest is the electricity used during corn chip manufacture (0.086 kg CO_2 eq per packet). Although the oil for frying the corn chips is the next single largest source (0.048 kg CO_2 eq per packet), the manufacture of the packaging (box plus packet, being 0.06 kg CO_2 eq) exceeds its greenhouse gas emissions. By sector, 6% of emissions are pre-farm, 36% are on-farm and 58% are post-farm (see Fig. 9.2).

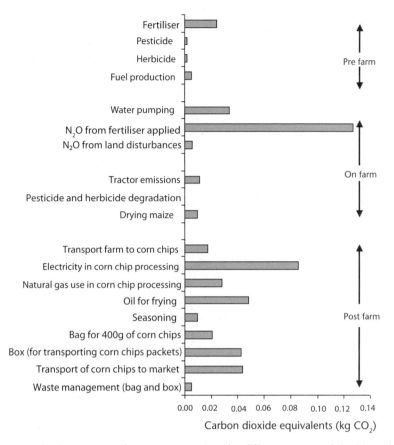

Figure 9.2 Contributions to greenhouse gas emissions for different stages of the life cycle of corn chips per 400 g packet of corn chips (after Grant and Beer 2008).

Although the basic life cycle results indicate that post-farm activities are more significant overall than on-farm activities and pre-farm inputs, nitrogen fertiliser application is the largest single contributor to greenhouse emissions. Moreover, if soil carbon is included (soil carbon and carbon dioxide emissions from soils are excluded in the main analysis because they are not counted as a greenhouse gas in national and international greenhouse gas inventories), measurements conducted throughout the project indicate that maize produced using stubble burning would rise to over 50% of total life cycle emissions (Kirkby *et al.* 2006). Post-farm, corn chip production is the single largest source of greenhouse gas emissions, although packaging and transport are also substantial, comprising 24% of total life cycle emissions, with packaging the third largest emission source overall.

Life cycle costs were obtained through the stakeholder surveys with local maize farmers and the corn chip manufacturer. The resultant relationship between the value chain and cumulative emissions is illustrated in Figure 9.3, showing the cumulative greenhouse gas emissions at each stage of the corn chip manufacture, along with an estimate of the appropriate value (based on the costs involved) added at each step of the production chain. The changing gradient indicates that pre-farm emissions add less value and generate more emissions proportionally per unit value added than post-farm activities. Indeed, pre-farm and on-farm operations add A\$0.4 value per kg of CO_2 eq greenhouse gas emitted and this figure rises to A\$2 value per kg of CO_2 eq greenhouse gas emitted for post-farm activities. Incidentally, this

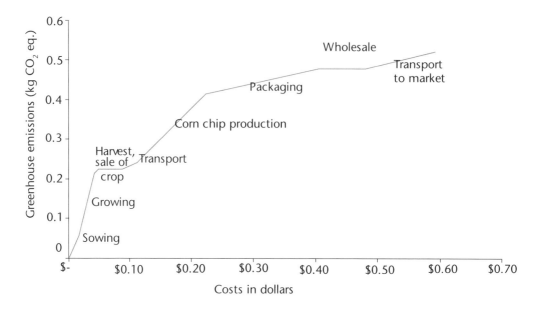

Figure 9.3   Cumulative greenhouse gas emissions and cost estimates for a 400 g packet of corn chips (after Grant and Beer 2008).

pattern could be even more marked for other maize products where the post-farm activities are less greenhouse intensive. For example, while in corn chip production 36% of greenhouse gas emissions are attributable before the farm gate, for ethanol production from maize the figure is 51%, and for starch production it is 58% (Fig. 9.3).

The study reveals important information for policy makers regarding on-farm emissions, emissions beyond the farm gate, and the relationship between these two parts of the system of corn chip production. On the farm, the pumping of water represents a significant energy use and consequent greenhouse contribution, particularly if the water is being drawn from deep bores – which may become a more prevalent practice with further water supply constraints. The study suggests that pumping irrigation water from deep wells causes almost three times more greenhouse gas emissions than irrigation using surface waters. Also, there appears to be substantial potential to reduce emissions on-farm through stubble management, although soil carbon and greenhouse gas emissions may also be influenced by other factors such as soil tillage, fertiliser and crop rotation regimes (Follett 2001; West and Marland 2002; West and Post 2002; Tilman *et al*. 2002; Duxbury 2005; Eckard 2006).

The substantial post-farm emissions indicate that there is also scope for exploring practical mitigation options, particularly in relation to transport and packaging. Moreover, the clear implication from the value chain results is that farming provides significant greenhouse gas emissions but less value-added, and therefore limited ability to invest in, greenhouse reduction innovations and practices. There would appear to be a *prima facie* case here for directing resources towards these greenhouse-intensive but relatively cash-poor parts of the production chain.

9.6   Miles to market: the food miles debate

Fresh produce has been travelling further and further to market over recent decades. One United Kingdom ((UK) study states that the amount of food moved by heavy goods vehicles has risen by 23% since 1978, and the average distance for each trip has increased by over 50%

(AEA 2005). The 'food miles' idea has found its potency in the 'local versus global' debate, which is prevalent in discussions regarding food and agriculture and beyond. The essential thesis is that food produced locally is better for the environment than that produced remotely and transported to market. This idea has been promoted by agricultural products producers who are under pressure from increasing imports, and from regional producers keen to establish points of differentiation between their products and more generic, distant offerings in the marketplace. Some environmental campaigners have also linked food miles concepts to 'anti-industrial', 'anti-globalisation' and 'anti-corporate' platforms.

All other things being equal, it is indeed true that food shipped using fossil fuels over low or no distance is likely to have a lower impact than food shipped using fossil fuels for long distances. However, there are two immediate problems with this starting point. First, all other things are invariably not equal, especially in the case of food production. Second, the implicit assumption in this focus on transport is that the transport part of the system dominates environmental impact. This is generally not the case in agricultural/food production systems.

The fundamental problem with the 'food miles' concept from an LCA perspective is that there is no logical environmental rationale for drawing the system boundary in such a way as to exclude all other process steps and activities except transport to market. As we have already identified in this chapter, on-farm activities and inputs are typically responsible for large portions of the impact burden for typical foods consumed in western countries. This is not to say that transport may not form a significant part of the overall impact burden, particularly where refrigeration and/or aviation are significantly involved, or where small volumes of high value food are transported long distances in complex logistics chains. However, to focus solely on transport is an immediate and clear error from a life cycle perspective. The major UK study on food miles puts the matter succinctly in the first recommendation of the report: 'A single indicator based on total food kilometres is an inadequate indicator of sustainability' (AEA 2005).

The DEFRA study (AEA 2005) also provides case studies suggesting that tomatoes grown in Spain may be significantly less greenhouse gas-intensive than tomatoes grown in gas-heated greenhouses in the United Kingdom, and that for chicken meat, the mode of processing and cooking may be the most significant single factor rather than transport. While the system boundary is truncated in these short case studies, they nevertheless illustrate the danger of using food miles as a broader environmental impact proxy. This work was conducted following rising concerns about food transport concerns in the UK – concerns that inevitably led to scrutiny of food imports from across the globe, including Australia and New Zealand. In particular, the idea that products such as apples and lamb should be imported from New Zealand was questioned because these are 'traditionally' produced domestically in the United Kingdom. A subsequent study undertaken by staff at Lincoln University in New Zealand summarises the 'food miles' debate, and compares more complete energy analyses for food production in Europe and New Zealand, and concludes that the globally traded products are generally responsible for lower greenhouse gas emissions than locally grown foods in Europe (Saunders et al. 2006). Key issues identified include: chilling requirements for extended storage of apples (since hemispheric seasons are alternate, in-season global transport may be more efficient than domestic production with cross-season storage); and fewer fertilisers may be required in New Zealand, where year-round grazing is also available, avoiding a heavy reliance on supplementary feeds.

Another important shortcoming of the food miles concept is that there is no explicit recognition that different transport methods generate different levels of impact. In fact, ambient shipping by sea is relatively highly energy efficient, and thus attracts very small proportions of a typical life cycle energy demand profile for agricultural products using this transport method in the production system. In contrast, extended chilled transport by road, or, more significantly, any bulk transport by air, will invariably multiply the proportion of energy demand attributable to transport.

Recognising that food miles is an inadequate proxy for environmental burden identification does not mean that the alleged benefits of local, organic, or home grown food should be rejected. Indeed, a UK study for the Department for Environment, Food and Rural Affairs (DEFRA: AEA 2005) includes a case study on imported organic wheat, which suggests that there are a wide range of benefits associated with organic production systems, including reduced greenhouse gas emissions, higher soil quality and biodiversity, and lower waste and pollution including eutrophication. The calculations presented indicate that energy savings in production of organic winter wheat may be equivalent to almost 800 kilometres of road transport. Logically, local, organic production may provide further benefits, provided these systems do not involve increased impacts elsewhere in the system. Anderson-Wilk (2007) quotes a report from the Iowa State University Leopold Centre for Sustainable Agriculture regarding 'community-supported agriculture' (CSA), which is based on local, communal gardening: 'CSA may minimize some of the negative effects of more conventional systems of food production and distribution because it involves less chemical use, less soil erosion, less food packaging, fewer food miles and more crop and ecosystem diversity' (Tegtmeier and Duffy 2005). There are many such statements in the literature, although the evidence in the form of systematic and complete LCA data and analysis is less widespread, and there is an urgent need for more studies indicating the full system benefits (and costs) of organic and localised production systems.

9.7 Discussion: LCA and farming in an environmentally constrained world

As our main food source, agriculture is a major, essential activity, and one which is contentious as a major user of land and water and emitter of greenhouse gases. This contention will increase as the remaining stock of 'natural' land is used, population and food consumption continues to rise, and scrutiny of sources of greenhouse gas emissions increases. Regarding the latter, agriculture is seen as both a major 'problem' (emitter) and potential 'saviour' – a source of fossil fuel substitutes and carbon sinks. The choices made about future technologies and practices will determine environmental outcomes, and LCA can play a key role in identifying preferable courses of action. Yet there are challenges to be overcome both in LCA practice and in realising sustainable agriculture options. Three key questions are posed here.

9.7.1 Can the 'new' agriculture be sustainable?

The industrial economy of the last 100 years has been based on historical biomass deposits in the form of crude oil, natural gas and coal. It is not surprising then that, as these historical reserves become limited in supply, society will look to more immediate biomass production systems to fill the gap (i.e. agriculture and forestry). Hence, the prospect of a 'new' agriculture industry is raised in which biomaterials, biofuels and other bio-energy sources will be increasingly produced, harvested and exploited in place of current fossil fuel-dependent technologies. Life cycle thinking, of course, cautions against the immediate conclusion that such technologies will be 'carbon neutral', particularly when we start to construct life cycle process chains of likely comparative systems. Fossil-based polymers or fuels have short production cycles based around winning crude oil and processing this into the final product, whereas agricultural alternatives often involve a long range of activities from seed manufacture to fertilisers, pesticides, tillages and harvesting, and onto processing and production.

A full discussion of the life cycle implications of the myriad of new bio-based technologies is outside the scope of this book, although discussion of biofuels is included in Chapter 10, and suffice to say, LCA is a useful tool for the comparison of different technological options and systems for producing similar products and services. Indeed, a not unrelated debate is that between 'natural' and 'synthetic' materials and products elsewhere in the economic system. LCA has contributed to debunking the myths around materials such as wool and nylon that 'natural' is always best. Semantically, while agriculture provides 'natural' alternatives to nylon, vinyl and other hydrocarbon based compounds, it is also the provider of 'synthetic' versions of various fuels and oils. Hence, the terms are essentially cultural, complex and contextual, and LCA provides a systematic means by which different delivery systems for similar services can be evaluated in environmental terms. The same applies to organic, permaculture, urban agriculture, genetically modified (GM), nanotechnologies, and so on; if data is available and impacts can be modelled within uncertainty limits, then LCA is appropriate.

From biodegradable shopping bags to bio-ethanol, agriculture is likely to be increasingly involved in the transition to lower fossil fuel dependency. The challenge for LCA is to describe transparently and communicate efficiently the relative impacts of different options in generalisable terms, to enable effective and appropriate decision-making concerning these new industries, and to enable the elimination of high-risk, unknown or relatively high impact options in favour of more benign ones.

9.7.2   What are the key constraints on LCA application to agricultural systems and how may they be overcome?

The diversity and openness of agricultural systems combined with gaps in data present ongoing challenges for LCA. For example, the IPCC notes that climate change impacts on agricultural pests, diseases, crop growth rates and yields, and water availability, are poorly understood (IPCC 2007). Furthermore, methodological and boundary issues remain. Time boundaries must be drawn carefully to accurately allocate impact. For example, arable crops are often grown in rotation and fertiliser applications, and soil nitrification processes link temporally to both previous and future crops. It has therefore been suggested that impacts should be allocated according to uptake and efficiency per crop, allocating impacts over successive years of different crops, using cropping plans and developed models of application/uptake of fertilisers and soil dynamics (van Zeijts *et al.* 1999). Another 'timing' issue relates to land use change. As indicated above, this is the single largest greenhouse impact of global agriculture, yet as in the sugar cane case, most agriculture LCA starts from a position of established farming systems. Since sugar cane in Queensland is long established, LCA does not typically include consideration of 'one-off' environmental burdens associated with changes in land use from pre-existing (natural) systems when this change took place in the past.

Timing is also related to our generally poor understanding of nutrient cycles in agricultural systems. Agricultural impacts can often hinge on a few poorly characterised inventory entries: effects of land clearing, fuel use on the farm, fertiliser, water, or N_2O emissions. The N_2O emissions are particularly sensitive due to their potency in climate change (Ehhalt and Prather 2001; IPCC 2006). However, our understanding of the relationship between nitrogen fertiliser application on crops and resultant fractional increases in soil N_2O emissions remains relatively rudimentary. In one study, the global climate change impact of extra N_2O entering the atmosphere as a result of producing biofuels crops is estimated to equal or exceed the benefits of this practice in substituted fossil fuels (Crutzen *et al.* 2007), thus questioning the entire case for biofuels agriculture (see also Chapter 10). Moreover, this discussion implies a focus on greenhouse gas emissions, whereas there are much greater gaps in data availability regarding other impacts of farming – water, biodiversity, land use and pollution, both in terms of inventory data and impact factors. This relates to the need for an 'eco-indicator' set for Australia as discussed elsewhere (see Chapter 5). Meanwhile, in the absence of reliable data and impact factors, a combination of caution and caveats are required so that the risks and uncertainties associated with poorly defined systems are adequately identified.

Variability between systems and farming conditions brings data transferability and methodology/boundary selection challenges. Hence, as in the sugar cane study, it may be difficult to draw 'average' conditions reliably and therefore make assumptions about system truncation. Consider the case of tomato ketchup. Tomato production systems vary according to availability of sunlight and growing temperatures. Also, while the production of the sauce is a more controlled and predictable process, there is then a less predictable and less known use phase. As highlighted in problems around Tesco's efforts to develop carbon labels in the UK, a bottle of ketchup that sits in a refrigerator for one year could be considered to contain 90% more embodied energy than a bottle that is consumed within one month.

It would be erroneous to conclude from all these difficulties that agriculture does not lend itself well to LCA. Although it is true that LCA was developed within the more controlled, predictable and readily measurable factory-industrial complex, and that farmland is infinitely variable, as is weather and climate, seed types, rates and practices of application of fertilisers and pesticides, proximity to and sensitivity of local hydrological systems and so on, LCA is the appropriate means to compare life cycle environment implications of different agricultural options. The answer is not to abandon LCA but to build more LCA capacity, to extend its use and redouble the research effort in agriculture – and to learn how to do LCA smarter. This can be achieved through further protocol and methods development and standardisation, to maximise comparability of studies and compatibility of data. It can also be achieved through the development of 'quick LCA' tools and calculators. These can be developed from better understanding of the variability of farming systems, and allow users to input various parameters to enable localised and specific evaluations to be achieved more quickly and easily than through full LCA studies.

9.7.3   What issues fall 'beyond' LCA and how can the interface between LCA and other design-decision support techniques be optimised?

Inevitably, there are challenges in the environmental assessment of agricultural practices for which LCA is unlikely to be the appropriate vehicle in finding the solution. For example, LCA is limited as a comparison tool in that it requires a similar unit of comparison to be achieved. Hence, comparing apples from an orchard system in Victoria with apples from an alternative system in New Zealand is within the bounds of LCA, provided we accept the assumption that these are the only important functional outputs of the two systems. In reality, agriculture often provides both economic and environmental/social services in addition to those associated with the food produced, such as rural landscape amenity value, natural heritage, ecosystem support and stewardship. Hence, it would be a mistake to advocate the adoption of LCA as the only or primary tool, as there will always be a need for tools other than LCA in environmental assessment of agricultural and related rural systems. Specifically, land management tools provide a site-specific context for study, whereas LCA is an essentially systematic technique, where assumptions need to be drawn across different situations to produce aggregated results.

Other challenges for LCA arise from dynamic economic factors that continually reshape agricultural systems. Population growth and agricultural technologies affect relative supply and demand and the debate over sustainable food supply is hundreds of years old and remains controversial. Bartle (2007) describes how agricultural production has outstripped population growth in many countries and the real price of food has dropped dramatically while food intake has increased. Across the European Union and the United States of America (USA), up to 20% of land has been set aside to avoid surplus agricultural production. However, the

United Nations Food and Agriculture Organization (FAO), in its projection for agriculture going forward to 2030 and 2050, suggests there could be a major food crisis looming: 'If no corrective action is taken, the target set by the World Food Summit in 1996 (that of halving the number of undernourished people by 2015) is not going to be met' (FAO 2008). An implication for LCA is that sustainability metrics that already have human health endpoints may need to incorporate effects of global supply chain on access to food, shelter and other basic needs.

Extending the debate further into the realm of social context, LCA studies need to inform and be informed by social factors. For example, even where LCA suggests that lower overall impacts may arise from a particular application of organic farming, this conclusion combined with a growing interest in organic and other less intensive agriculture may not be enough to change practice effectively and quickly. Habits, perception and social practice are linked to norms, values capacities, institutional frameworks and infrastructure, and a systemic focus on short-term production creates a blind spot over the need for long-term maintenance and stewardship (Hill 1998). The need for LCA in agriculture is perhaps greater than in other sectors, given the complexity, variability and lack of previous attention, and new LCA will be most effective in driving change when it is linked to applied socioeconomic assessment and research techniques to support policy development.

Economics also often dominates the 'gap' between LCA-based environmental optimisation and reality. The example of LCA of maize farming and corn chips production neatly illustrates a significant problem likely to be applicable across many agricultural systems; a significant contribution to the total greenhouse gas impacts occur before the farm gate where the economic value is low and the returns marginal, whereas most of the value addition occurs beyond the farm gate. Hence, in non-vertically integrated production systems, there is a mismatch between the capacity of the economic unit to respond to the need to reduce agricultural impact and the impact itself. Due to this economic situation and a range of other institutional and structural factors, the innovation capacity among agricultural producers is typically lower than in the 'value-adding' food processing industries.

In a future economic system where there is more attempt to internalise environmental impacts (e.g. through carbon trading or taxes), economic decisions may be made increasingly along LCA lines. The local implications could be socially and economically severe. For example, in Victoria, which is a significant exporter of dairy products high in embodied greenhouse gas emissions, there is an effective and significant export of greenhouse gases taking place. Economic forces may lead to a reshaping of this pattern, heralding a climate-optimised system of farming that is centred around rural carbon management rather than food production.

Key questions are: what is the limiting factor in the future world? Is it land, or greenhouse gas emissions or water? In Australia it is probably the latter two, but this may vary from region to region. LCA must therefore develop in order to be geographically and locationally specific enough to provide the appropriate answers to the appropriate (limiting) questions of environmental capacity and burden, and become increasingly sophisticated in linking to social, cultural and economic issues and the appropriate methods by which these are valued and described.

Given the dynamics and importance of agriculture and its impacts on the environment, there is a clear role for LCA in both 'conventional' and 'new' agriculture, providing as it does a powerful framework for assessing which of the many uses of biomass/crops lead to maximum benefits to society in terms of functionality and displacement of unsustainable practices. However, it is also clear that the bio-economy is not going to replace the fossil-based economy on a one-to-one ratio. Sharing the world's renewable production capacity among all human needs and wants is going to require restructuring the way we consume products and services, with implications beyond agriculture.

9.8 Conclusions

The case studies and discussion indicate that agricultural LCA is particularly important because there are significant impacts associated with land and water use, greenhouse gas and other pollutant emissions. In particular, the conversion of land to agriculture, N_2O emissions from soils, and methane emissions from livestock are major factors in climate change. LCA has already made significant contributions to sustainable agriculture, for example, in identifying 'counter-intuitive' results indicating that 'natural' products often cause more environment burdens than 'synthetic' versions, and that 'food miles' is not necessarily an appropriate environmental proxy. It follows that a strengthened and more extensive application of LCA may minimise the risk of going down such methodological cul-de-sacs in future.

Three trends will determine the extent of the future successful application of LCA in informing agricultural practice to minimise environmental impacts. First, LCA data and tools must be further developed to allow flexible, standardised, yet sufficiently accurate application of LCA at the farm/field scale with appropriate temporal specificity. As our understanding of key variables in agricultural systems develops, there is a role for LCA-based calculator tools to enable quick assessment of different options in the widely varying ranges of conditions across farming systems.

Second, agricultural stakeholders need to embrace LCA as a technique that can contribute to farm-scale decision-making, and contribute data and resources accordingly. This, of course, is linked to the availability of reliable and accessible LCA analyses, hence the need for calculators. Third, given the mismatch that can occur between capacity and environmental opportunity (e.g. in the corn chips case study), government and the policy-making community must contribute to the resourcing of LCA uptake in agricultural applications and recognise and utilise the results of LCA in developing policy initiatives to encourage environmentally sustainable agricultural systems. This requires LCA to become more sophisticated in linking with academic disciplines and other techniques that also contribute to sustainability assessment and decision support.

9.9 References

AEA (2005) 'The validity of food miles as an indicator of sustainable development.' Final Report Produced for UK Government Department for Environment, Farming and Rural Affairs (DEFRA). AEA Technology, Harwell, UK.

AGO (2003) 'Australian methodology for the estimation of greenhouse gas emissions and sinks 2003.' Agriculture. Australian Greenhouse Office, Department of the Environment and Heritage, Canberra.

Anderson-Wilk M (2007) Does community-supported agriculture support conservation? *Journal of Soil and Water Conservation* **62**(6), 126A.

Bartle (2007) The global agricultural surplus and the case for non-food crops. Retrieved 18 March 2008 from <http://www.futurefarmcrc.com.au/documents/Globalagricsurplus.pdf>. Future Farm Industries Cooperative Research Centre.

Beer T, Meyer M, Grant T, Russell K, Kirkby C, Chen D, Edis R, Lawson S, Weeks I, Galbally I, Fattore A, Smith D, Li Y, Wang G, Park KD, Turner D and Thacker J (2005) 'Life-cycle assessment of greenhouse gas emissions from agriculture in relation to marketing and regional development – irrigated maize: from maize field to grocery store.' Final Report HQ06A/6/F3.5, CSIRO Division of Marine and Atmospheric Research, Aspendale, Victoria.

Bellarby J, Foereid B, Hastings A and Smith P (2008) 'Cool farming: climate impacts of agriculture and mitigation potential'. Greenpeace International, Amsterdam.

Brentrup F, Küsters J, Lammel J and Kuhlmann H (2000) Methods to estimate on-field nitrogen emissions from crop production as an input to LCA studies in the agricultural sector. *International Journal of Life Cycle Assessment* **5**(6), 349–357.

Crutzen PJ, Mosier AR, Smith KA and Winiwarter W (2007) N_2O release from agro-fuel production negates global warming reduction by replacing fossil fuels. *Atmospheric Chemistry and Physics Discussions* **7**, 11 191–11 205.

Denmead O, MacDonald B, Bryant G, Reilly R, Griffith D, Stainlay W, White I and Melville M (2005) Gaseous nitrogen losses from acid sulfate sugarcane soils on the coastal lowlands. *Proceedings of the Australian Society of Sugar Cane Technologists, 3–6 May 2005, Bundaberg, QLD.* ASSCT: 211–219.

Duxbury JM (2005) *Soil Carbon Sequestration and Nitrogen Management for Greenhouse Gas Mitigation, in Climate Change and Agriculture: Promoting Practical and Profitable Responses.* p. IV-5–IV-7. <http://www.climateandfarming.org/pdfs/FactSheets/IV.2Soil.pdf>.

Eckard R (2006) Are there win-win strategies for minimising greenhouse gas emissions from agriculture? *Proceedings of OUTLOOK 2006*, Australian Bureau of Agricultural and Resource Economics (ABARE), Canberra.

Edis R, Chen D, Wang G, Turner D, Park K, Meyer M and Kirkby C (2008) Soil nitrogen dynamics in irrigated maize systems as impacted on by nitrogen and stubble management. *Australian Journal of Experimental Agriculture* **48**(3), 382–386.

Ehhalt D and Prather M (2001) Atmospheric chemistry and greenhouse gases. In: *Climate Change 2001: The Scientific Basis.* (Eds JT Houghton, Y Ding, DJ Griggs, M Noguer, PJ van der Linden, X Dai, K Maskell and CA Johnson) pp. 239–287. Cambridge University Press, Cambridge.

FAO (2008) UN Food and Agriculture Association website. Retrieved 3 March 2008 from <http://www.fao.org/es/ESD/gstudies.htm>.

Feitz AJ, Lundie S, Dennien G, Morain M and Jones M (2007) Generation of an industry-specific physico-chemical allocation matrix: application in the dairy industry and implications for systems analysis. *International Journal of Life Cycle Assessment* **12**(2), 109–117.

Flessa H, Ruser R, Dörsch P, Kamp T, Jimenez MA, Munch JC and Beese F (2002) Integrated evaluation of greenhouse gas emissions (CO_2, CH_4, N_2O) from two farming systems in southern Germany. *Agriculture, Ecosystems and Environment* **91**, 175–189.

Follett RF (2001) Soil management concepts and carbon sequestration in cropland soils. *Soil and Tillage Research* **61**, 77–92.

Global Footprint Network and the University of Sydney (2005) *The Ecological Footprint of Victoria Assessing Victoria's Demand on Nature.* EPA Victoria, Melbourne.

Grant T and Beer T (2008) Life cycle assessment of greenhouse gas emissions from irrigated maize and their significance in the value chain. *Australian Journal of Experimental Agriculture* **48**, 1–8.

Hill SB (1998) Redesigning agroecosystems for environmental sustainability: a deep systems approach. *Systems Research and Behavioural Science* **15**, 391–402.

IPCC (1997) *Greenhouse Gas Inventory Reporting Instructions.* Revised 1996 IPCC Guidelines for National Greenhouse Gas Inventories, Volumes 1–3. The Intergovernmental Panel on Climate Change, London, UK.

IPCC (2006) *2006 Guidelines for National Greenhouse Gas Inventories.* Volume 4, Chapter 11: N_2O emissions from managed soils, and CO_2 emissions from lime and urea application. Institute for Global Environmental Strategies (IGES), Hayama, Japan.

IPCC (2007) 'Climate change 2007: impacts, adaptation and vulnerability'. Contribution of Working Group II to the Fourth Assessment Report of the Intergovernmental Panel on Climate Change. p. 285. Cambridge University Press, Cambridge.

Kirkby C, Fattore A, Smith D and Meyer M (2006) Life cycle assessment of greenhouse gas emissions from irrigated maize stubble treatments and plant/soil responses. In: *Water to Gold, Proceedings of the Maize Association of Australia 6th Triennial Conference.* (Eds E Humphreys, K O'Keeffe, N Hutchins and R Gill) pp. 177–184. The Maize Association of Australia, Shepparton, Vic.

Meyer CP, Kirkby CS, Weeks I, Smith DJ, Lawson S, Fattore A and Turner D (2006) Nitrous oxide production from irrigated maize cropping in the Murrumbidgee Irrigation Area: impacts of crop residue management systems. In: *Water to Gold, Proceedings of the Maize Association of Australia 6th Triennial Conference.* (Eds E Humphreys, K O'Keeffe, N Hutchins and R Gill) pp. 161–167. The Maize Association of Australia, Shepparton, Vic.

Nicol R (2005) A life cycle view of Australian dairy products: lessons and opportunities from Dairy Australia Research. Presentation at 4th LCA Australia, Sydney.

Renouf M (2006) LCA of Queensland cane sugar – lessons for the application of LCA to cropping systems in Australia. *5th Australian Conference on Life Cycle Assessment.* Melbourne, 22–24 November. Australian Life Cycle Assessment Society, Melbourne.

Renouf M (2007) Comparing agricultural crops for bio-product applications. LCA in Food, Gothenburg, Sweden, 25–26 April. In SIK Annual Report 2007. Swedish Institute of Food and Biotechnology (SIK), Gothenburg.

Renouf M, Antony G and Wegener M (2005) Comparative environmental life cycle assessment (LCA) of organic and conventional sugar cane growing in Queensland. *27th Conference of the Australian Society of Sugar Cane Technologists (ASSCT).* Bundaberg, 3–6 May. ASSCT, Mackay, Queensland.

Renouf MA and Wegener MK (2007) Environmental life cycle assessment (LCA) of sugar cane production and processing in Australia. *Proceedings of the Australian Society of Sugar Cane Technologists* **29**, CD-ROM.

Saunders C, Barber A and Taylor G (2006) 'Food miles – comparative energy/emissions performance of New Zealand's agriculture industry.' Research Report No. 285, AERU, Lincoln University, July.

SoE (2006) 'State of the Environment 2006'. Department of the Environment and Heritage, Canberra.

Tegtmeier E and Dufiy M (2005) Community Supported Agriculture (CSA) in the Midwest United States. Leopold Center for Sustainable Agriculture, Iowa State University, Ames, IA.

Thorburn PJ, Horan HL and Biggs JS (2004) The impact of trash management on sugar cane production and nitrogen management: a simulation study. *Proceedings of the Australian Society of Sugar Cane Technologists* **26**, CD-ROM.

Thorburn PJ, Meier EA and Probert ME (2005) Modelling nitrogen dynamics in sugar cane systems: recent advances and applications. *Field Crops Research* **92**, 337–351.

Tilman D, Cassman KG, Matson PA, Naylor R and Polasky S (2002) Agricultural sustainability and intensive production practices. *Nature* **418**, 671–677.

US EPA (2006) Global Anthropogenic non-CO_2 greenhouse gas emissions: 1990–2020. 430-R-06-005, US Environmental Protection Agency, Washingon DC.

van Zeijts H, Leneman H and Wegener Sleeswijk A (1999) Fitting fertilisation in LCA: allocation to crops in a cropping plan. *Journal of Cleaner Production* **7**, 69–74.

West TO and Marland G (2002) A synthesis of carbon sequestration, carbon emissions, and net carbon flux in agriculture: comparing tillage practices in the United States. *Agriculture, Ecosystems and Environment* **91**, 217–232.

West TO and Post WM (2002) Soil organic carbon sequestration rates by tillage and crop rotation: A global data analysis. *Soil Science Society of America Journal* **66**, 1930–1946.

Wood R, Lenzen M, Dey C and Lundie S (2006) A comparative study of some environmental impacts of conventional and organic farming in Australia. *Agricultural Systems* **89**, 324–348.

Chapter 10

Climate change responses: carbon offsets, biofuels and the life cycle assessment contribution

Scott McAlister and Ralph E Horne

Carbon accounting has rapidly gained prominence because of the increasing recognition of the need to reduce greenhouse gas emissions as a result of evidence from the Intergovernmental Panel on Climate Change (IPCC 2007a, b, c) and the Stern Review for the British government (Stern 2006), among other authorities. Any scheme to reduce greenhouse gas emissions in a systematic way, such as the Kyoto Protocol or regional schemes, requires a standard methodology to account for the generation and sequestration of emissions. This is also a requirement for carbon trading schemes and carbon taxes.

LCA has a solid historical connection with carbon accounting since its past lies at least partly in energy accounting – and carbon emissions are dominated by the burning of fossil fuels for primary energy supply. Various attempts have been made to adapt LCA techniques to calculating the net carbon benefits of offsets and substitution of biomass. Initiatives have sought to standardise methods and approaches through programs such as the International Energy Agency (IEA) Task 38 and the European Commission-funded BIOmass-based climate change MITigation through Renewable Energy (BIOMITRE) project (Horne and Matthews 2004; van Dam *et al.* 2004).

In this chapter, the practice of carbon accounting is outlined and current controversies are investigated. A short description of existing standards for carbon accounting is presented, followed by a review of carbon accounting, life cycle assessment (LCA) and related methods. The remainder of the discussion focuses on three main case studies. The first is the use of forestry schemes as carbon offsets to compensate for greenhouse gas emissions associated with other activities. The second is the use of biomass energy technologies in place of fossil fuel-based energy technologies. The third illustrates LCA's potential role in and contributions to the uptake of carbon accounting.

10.1 Carbon accounting standards and assessment tools

There are two standards commonly used in carbon accounting: the international standard from the International Organization for Standardization's (ISO) ISO 14064 (ISO 2006a, b, c), and the World Resources Institute and World Business Council on Sustainable Development (WBCSD) 'Greenhouse Gas Protocol' (WRI-WBCSD 2004, 2005). ISO 14064 is divided into three parts, while the Greenhouse Gas Protocol consists primarily of two separate but linked standards; the first dealing with corporate level accounting and the second for project level accounting. Both standards are similar in intent and content and are complementary. ISO 14064 is less descriptive and shorter than the Greenhouse Gas Protocol, while the Greenhouse

Gas Protocol in turn provides a more ambitious approach. If one standard is met, the other is also likely to be met.

The following elements are common to both ISO 14064 and the Greenhouse Gas Protocol:

- determining the boundaries for greenhouse gas accounting – the Greenhouse Gas Protocol gives more guidance and examples than ISO 14064
- classification of emissions – although named differently in ISO 14064, the classifications are broadly similar to those provided by the Greenhouse Gas Protocol with more guidance
- identification and calculation of greenhouse gas emissions
- rules for changing base year inventories
- rules for tracking emissions over time
- rules for assessing uncertainty
- rules for greenhouse gas reporting.

Given the urgency of the need for effective carbon management, both the presence of carbon accounting standards and the methodological history of LCA are considerable assets. In Australia's National Greenhouse Factors report (Department of Climate Change 2008), based on the Greenhouse Gas Protocol, emissions are classified as:

- scope 1, or direct emissions from within the boundary of an organisation such as fuel combustion
- scope 2, or indirect emissions from the consumption of purchased electricity
- scope 3, which includes other indirect emissions that are a consequence of an organisation's activities, such as the extraction and transport of fuels included in scopes 1 and 2, losses in the delivery of electricity, travel and transportation of products, and any other products or services consumed by an organisation.

Users of this framework have difficulty deciding which scope 3 emissions to include in reporting, beyond those associated with the delivery of fuels. For example, for an employee embarking on air travel for business, the emissions are generated in the course of making a financial profit and therefore, it can be argued, are just as important as electricity used directly to manufacture a product. However, the emissions generated by the flight are the responsibility of the airline operating the service, as they are covered under scope 1 emissions for the airline. These difficulties in deciding what to include and exclude under scope 3 emissions can lead to quite different total carbon emissions – and also potentially to double counting. However, LCA can solve this dilemma by first providing a functional basis to the measurement. The extent to which things are included depends on their connection to the functional delivery. The issue of double counting is avoided, as emissions are followed along supply chains. Each part of the chain includes all emissions up to that point in the supply chain. Each step includes its 'own' emissions and those of all suppliers. The Goods and Services Tax (GST) system in most companies works on this principle for calculating taxes payable at any given point in the chain. Each company only effectively pays for the 'value addition' they make to the product or service, but they are also responsible for data collection regarding the tax from their inputs. The benefits of this approach are that shared responsibility for emissions provides improved data quality throughout the chain.

Greenhouse accounting systems provide two distinct reporting approaches for organisations with joint operations: equity share and control. In the equity share approach, emissions counted by a company are based on the company's equity share in the operation under investigation, which will normally be the percentage ownership of the operation. In the control approach, a company accounts for 100% of emissions from operations over which it has either financial or operational control, but it does not count those operations in which it has an equity share but no control. These different approaches to determining organisational boundaries can produce different results. While the difference is less important for a company voluntarily reporting its greenhouse gas emissions, the determination of boundaries can give a company a financial advantage or disadvantage in a carbon trading or tax scheme. Also, importantly, there is potential for double counting if two or more companies hold interests in the same organisation but use different reporting approaches. A functional supply chain approach would avoid these issues, as the eventual impacts of carbon emissions would belong appropriately to end consumers of products and services. The reluctance to use this approach may be linked to the apportionment of blame for emissions. Consumers are reluctant to accept responsibility for emissions along the supply chains that produce their goods and services. It may also serve economic interests to encourage an accounting approach where technical solutions rather than changes in consumption patterns are promoted as 'the way' to tackle greenhouse emissions – despite the constraints inherent in this approach.

These apparent complexities involved in greenhouse gas assessment necessitate clear rules and accounting systems. Accordingly, in Australia, demand for options to allow companies to purchase offsets and become 'carbon neutral' has led to a specific standard and certification process. The Greenhouse Friendly™ certification standard aligns with the Greenhouse Gas Protocol (AGO, 2006; Department of Climate Change, 2008). Greenhouse Friendly™ was launched in 2001 and has since been expanded as part of the Australian Government's Greenhouse Challenge Plus program. Under the program, credits from abatement projects approved by the Australian Government under the Greenhouse Friendly™ initiative are available for purchase by those seeking approved carbon offsets.

10.2   Carbon offsets: assessment problems associated with bio-sequestration

Bio-sequestration is being used increasingly to offset carbon emissions. By planting trees and other crops, organisations can 'neutralise' their emissions with the assumption that the emissions will be sequestered in plants as carbon. This approach, however, presupposes a sound knowledge of biomass cycles in the land and plants that are offset. These cycles are complex, however, as indicated even in the simplified schematic in Figure 10.1. Accordingly, carbon accounting in agriculture and forestry is plagued with high levels of multiple uncertainties.

A simple description of the generic bio-sequestration system illustrates the key uncertainties in predicting sequestration offsets. Theoretically, the release of carbon dioxide (CO_2) through burning fossil fuels is counterbalanced by the take-up of equivalent carbon through plant growth under the sequestration program. The first assumption here is that the plants in question would not be grown if not for the sequestration program and, furthermore, that the land upon which they are grown would not otherwise fix any carbon (carbon dioxide is referred to here in shorthand as 'carbon'). It is problematic to argue that fallow land will remain fallow and incapable of fixing carbon without a sequestration planting. Also, any fossil fuel consumption or soil carbon losses involved in preparing land and growing plants should also be counted.

It is also assumed that carbon dioxide absorbed during plant growth is 'locked in' for a significant period of time. Indeed, sequestration schemes generally take a land use change approach, which assumes that the sequestration crop will comprise the new land use in perpetuity, so that over the cycles of plant growth and harvests an average 'stock' of carbon will be maintained. Another key assumption is that the greenhouse gas balance of the soil and substrate remain constant with no net emissions of greenhouse gases. However, the mobilisation of soil carbon and other greenhouse gases is complex and the balance may not be constant.

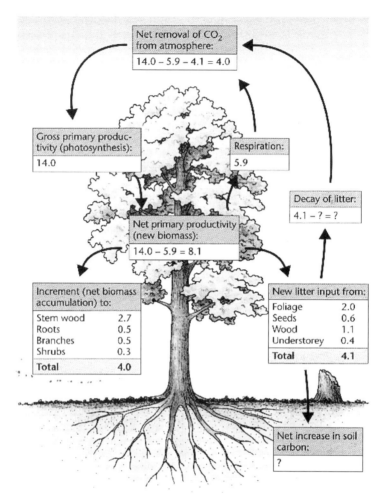

Figure 10.1   Schematic of forest biomass circulation system (numerals show tonnes of carbon per hectare per year) (cited in Horne and Matthews 2004).

Soils release various gases, particularly when disturbed or the overlying canopy and plant matter is changed, and these can include potent greenhouse gases such as nitrous oxide (N_2O) and methane (CH_4) in addition to carbon dioxide. Of course, a shift from agriculture to forestry may lead to positive greenhouse gas changes, as forests involve less soil disturbance than agricultural land.

Time scales, land use prior to and after planting the sequestration crop, the fate of the biomass after harvesting and soil dynamics are all potentially important factors in the calculation of 'deemed' net carbon emission reductions or offsets. Clearly, carbon accounting of bio-sequestration projects is contentious and there are various factors which should be taken into account in deeming any carbon offsets. In the following discussion, the example of forestry is used, although the main discussion points are also likely to apply to other forms of bio-sequestration.

10.2.1   Forests as carbon sinks

Forests are in a state of continuous carbon flux. During natural growth and regeneration, equilibrium is maintained through the carbon dioxide cycle – fixed by plant growth, emitted to air, locked up in soil and removed through weathering. Following forest harvesting and subsequent replanting, there is an abrupt adjustment to the balance, and forests become net emitters of atmospheric carbon for up to 20 or even 40 years (Marland and Schlamadinger 1997; Reijnders and Huijbregts 2003). Trees only become carbon 'sinks' when they are well-established. There are several sources of carbon emissions after harvesting. The roots and the trunk parts that remain after cutting can range from 15% to 40% of the total tree biomass, while tops, branches and bark residues typically account for another 15% to 20% of the biomass (Fraanje and Lafleur 1994; Reijnders and Huijbregts 2003). Therefore, 30% to 60% of total biomass can remain in the forest, where it degrades to form carbon-containing gases, the exact amount depending on harvesting practices such as the removal of branches. In addition, carbon contained in the soil is released due to soil disturbance associated with harvesting (Reijnders and Huijbregts 2003).

Forests do, of course, absorb atmospheric carbon and store it in solid wood, vegetation, litter and soil (Oneil *et al.* 2007). Carbon accumulation is typically linear in time during the initial phases of growth until half the maximum carbon sequestration has occurred, at which time accumulation slows and progresses asymptotically (without ever being completed) (Marland and Schlamadinger 1997). Different species of trees, as well as localised soil and climate conditions, also affect growth rates. Therefore, the age of a forest, its tree species and localised conditions will all have impacts on total carbon sequestration rates.

Land use prior to and following a forestry plantation can be expected to have a major role in net sequestration, as different land uses result in different soil carbon levels. If the previous land use is long-term forestry, the carbon level in the soil will be similar after afforestation. However, if the previous land use is agriculture, forestry will typically be expected to lead to a net increase in soil carbon (Reijnders and Huijbregts 2003). Similarly, where post-harvest land use introduces a change to agricultural land, the amount of carbon sequestered in the soil is likely to fall due to soil disturbances and adjustment to a new lower carbon soil regime (Reijnders and Huijbregts 2003). Moreover, as already mentioned, levels of other more potent greenhouse gases such as nitrous oxide can be important and should be included in calculations. Moisture content can locally determine nitrous oxide emissions, which can vary between fields as well as between regions (e.g. Beer *et al.* 2005; see also Chapter 9). This further complicates the problem of identifying the actual 'savings', and accounting systems allow for aggregate calculations of these based on standard conditions.

As indicated above, the fate of the harvested biomass also needs to be taken into account. Durable forestry products such as building materials store embodied carbon over their functional lifetime. At the end of their life, however, some of the embodied carbon is released, although rates of release depend on how products are disposed. Burning and aerobic composting will release carbon rapidly, while burial in landfill sequesters more carbon for longer and leads to conversion of the remaining biomass to carbon dioxide and methane. Shorter term products such as paper will typically be burned (emitting carbon dioxide) or buried in landfill where they will decompose more quickly and more thoroughly than durable forestry products in landfill – emitting carbon dioxide and methane. Depending on the landfill technology, more than half the methane will be captured for power generation or flared to reduce safety risk and greenhouse potential.

Clearly, time – including forest life span and age – also affects total carbon sequestration potential. However, this needs qualification as, again, there are complicating variables. Older forests in Australia generally present increased risk of wildfire from a combination of natural mortality and expected increases in temperature due to global climate change (Gedalof *et al.* 2005; Oneil *et al.* 2007). Climate change can also be expected to increase the risk of insect and disease outbreaks in forests, increasing mortality and hence the risk of wildfire (Oneil *et al.* 2007). Wildfires release stored carbon, turning forests from carbon sinks into carbon sources (Westerling *et al.* 2006; Oneil *et al.* 2007).

While forests and tree planting are almost universally lauded for their positive impacts on species diversity, carbon capture and climate regulation, a life cycle perspective is still required to look at the systemic issues involved in increasing forests and tree planting. Young forestry trees in particular consume large quantities of water. This potentially reduces water flows to both surface and groundwater systems, depending on local conditions. As many areas in the world are expected to become water-stressed due to climate change, increased forestry for sequestration could reduce water available for other uses, including environmental flows. The National Water Initiative (Council of Australian Governments 2005) indentified large-scale forestry as having the potential to intercept large volumes of surface and groundwater flows. This will affect both economic and environmental uses of downstream water flows from such projects. This issue is being incorporated only recently into some planning for tree planting for carbon capture.

Similarly, land planted for carbon sequestration means it is not available for other uses. This may reduce greenhouse gas emissions (e.g. when the carbon content of soil is increased due to a change from agriculture to forestry). However, it may also mean that a monoculture of a single tree species is created in place of a previously rich ecosystem of multiple species, thus reducing biodiversity.

Notwithstanding the uncertainties and difficulties associated with assessment of the relative value of different forests as carbon sinks, both LCA and greenhouse standards are being applied successfully. For example, Greenfleet, a relatively long-established not-for-profit provider of offsets specifically aimed at fossil-based transport users and providers, is certified under the Greenhouse Friendly™ scheme in Australia (see above) to provide high-quality abatement offsets.

10.3   Low carbon energy: LCA and biomass technologies

Biomass energy technologies being developed are invariably intended to substitute for existing fossil fuel-based energy systems (Fig. 10.2). Biomass technologies may be simpler and more reliable options than other offsets in particular because the timeframe involved in achieving the abatement is essentially the time up to harvest and utilisation. Indeed, the burning of biomass as a substitute for the burning of fossil fuels is widely advocated as a method to reduce greenhouse gas emissions. However, there are a range of concerns about biofuels and other biomass energy products. For example, they invariably involve fossil fuel consumption in their production and processing and also often require land that could otherwise be used for food production or maintained as biodiversity reserves.

Clearly, it is a mistake to assume that burning biomass is always greenhouse-neutral or desirable just because carbon dioxide emissions from burning biomass are biogenic and part of the short-term carbon cycle. Although it is true that carbon dioxide emitted in burning is originally sequestered from the atmosphere when the biomass grows, and therefore overall no net carbon dioxide emissions occur, there are other greenhouse gases emitted from biomass energy technologies.

The IPCC considers that biogenic emissions do not add to global greenhouse gas emissions, and they are therefore not taken into account in most regulatory and accounting schemes. There is, however, uncertainty in determining the savings achievable by substituting biomass for fossil fuels. This is an important consideration in the context of an emissions trading scheme, where a reduction in carbon dioxide emissions results in monetary benefits. The following case study illustrates some issues associated with calculating reduction in carbon dioxide emissions associated with biomass substitution in energy generation. Similar discussion points are relevant for any biomass substitution scenario.

10.3.1   Using biomass in energy generation

As with any LCA, the first consideration in determining potential savings from a biomass energy technology is the analysis boundary. A simple boundary condition would be greenhouse gas emissions solely at point-of-use, counting net emissions from the biomass upon combustion and subtracting emissions due to avoided fossil fuel combustion, to provide an overall indication of the greenhouse benefit of the technology. However, this does not necessarily provide a realistic approximation of net greenhouse gas emissions from the combustion of biomass. The system boundary should be extended to include inputs from growing, harvesting and transporting the biomass, and the contribution of soil carbon sequestration. This extension should also include a consideration of the 'upstream' fossil fuels system, including mining, processing and transporting fossil fuels. The greater the level of extension, the more comprehensive the consideration of greenhouse gas emissions from the biomass technology, although as the boundary expands so does the level of data and uncertainty in assumptions about these expanded elements.

For example, fertiliser, herbicide and pesticide production all contribute to greenhouse gas emissions associated with (non-organic) biomass crops. There are variations in both the type and quantities of inputs applied to a given quantity of biomass. Practices vary, local conditions and soil optimisation requirements vary, and emissions factors vary between electricity systems within which these products are manufactured. Further uncertainties occur for short rotation crops, since a percentage of fertiliser applied at a given time may be for the current crop to use although a further percentage may be for use by a subsequent crop, and therefore it needs to be allocated.

There is also uncertainty associated with changes in carbon storage in soils and soil-related greenhouse gas sources, as discussed in Section 10.2. This is especially important, as these effects may be significant in the total greenhouse gas balances of biofuels. For example, where switchgrass was substituted for coal in an existing coal-fired electricity generation facility, soil carbon sequestration accounted for 29% of the net benefit from the fuel switch, but it contributed 62.5% of the total uncertainty (Ney and Schnoor 2002).

Another important consideration in calculating greenhouse gas balances for biomass energy technologies relates to the specific substitution situation. In the case of local off-grid co-firing, the calculation will be based on the primary fuel or mix of fuels being replaced. In the case of a stand-alone biomass plant that contributes to a region or country's electricity grid, the calculation of greenhouse benefits is more complex. Arguably, it is inappropriate to select an expensive, unused, inefficient or obsolete fossil fuel technology for comparison in calculating net benefits. Logically, the appropriate comparison is with a fossil fuel technology that could reasonably be adopted given the market conditions and current technologies in the selected setting. One study (Schlamadinger *et al.* 1997) suggests that a reference energy system should be 'the least-cost fossil energy system with the lowest greenhouse gas emissions and minimised environmental impacts, fulfilling the same goals as the bioenergy system'.

Another approach is to compare the marginal electricity generator, provided the biomass energy alternative can meet marginal demand requirements (i.e. it can be easily and quickly brought on-stream at energy demand peaks and shut down following them). For example, in a coal-dominated electricity grid, the marginal electricity generator might be a gas-fired plant. The addition of new capacity from a biomass project would displace the gas-fired plant, and therefore gas rather than coal, or a mix of gas and coal, would be the fossil fuel being replaced. Where electricity demand exceeds supply, the addition of the biomass project may increase overall energy consumption rather than displace an existing fuel. In this case, greenhouse gas emissions would not necessarily be reduced, and the only benefit of using biomass technology

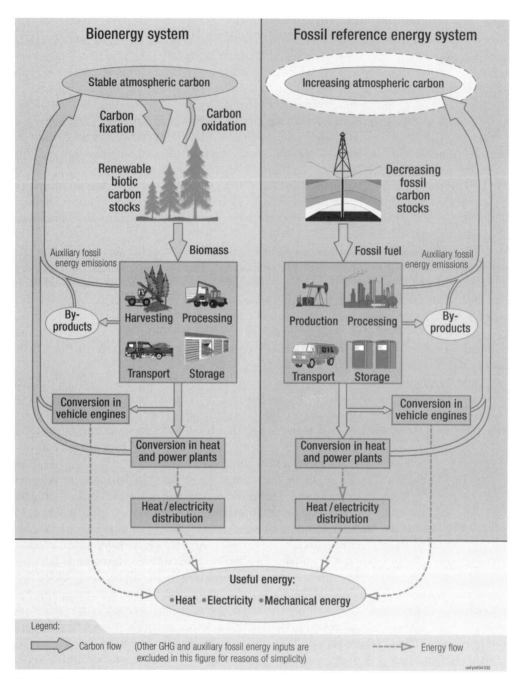

Figure 10.2 Comparison of bioenergy and fossil energy systems (cited in Horne and Matthews 2004).

would be the marginal emission increase foregone through selecting the biomass technology rather than a fossil alternative.

The system boundary of comparative assessment may also be extended further in cases where waste biomass is the main biomass source. In such cases, the fate of the waste biomass in the default (fossil fuel) scenario should be considered in addition to the credits due to fuel substitution. Hence, some authors argue that the use of biomass waste or residue in power production should be credited with the savings in emissions associated with the normal route of biomass disposal in addition to the savings due to forgone fossil fuel combustion (Mann and Spath 2001). This may add credit to the use of waste biomass, since typical biomass disposal often contributes to methane emissions.

Lastly, it has been argued (Reijnders and Huijbregts 2003) that because forests sequester carbon dioxide irrespective of whether it is biogenic or fossil fuel-derived, the carbon contribution of biomass burnt for electricity generation should be counted as if it was fossil fuel-derived. Accordingly, as 14% of worldwide carbon dioxide emissions are currently sequestered, 14% of biomass burnt should be considered as sequestered, while the remaining 86% should be counted in overall emissions because burning biomass directly contributes to an increase in atmospheric carbon dioxide.

Across this range of approaches and options for greenhouse emission abatement, the approach outlined in the methods chapter (see Chapter 2) regarding consequential LCA analysis provides a clear framework for assessing the change in systems resulting from specific decisions. These decisions may be in the product or process selection, or financial investment in the form of carbon trading, and thus lead to environmentally beneficial changes in practice being accelerated through the economy. If an LCA timeframe is chosen, at least 500 years or more, any short-term issues surrounding land use equilibrium, or building product use and disposal are resolved.

10.3.2   Biomass carbon assessment tools

Given the varying complexity, the methodological challenges and the high level of interest in biomass energy technologies and bio-sequestration, it is not surprising that several methodologies and tools have been developed for evaluating the greenhouse gas balances and cost-effectiveness of various biomass energy technologies. There is no standard tool, and typically each is proprietary, complex and/or restricted to one biomass energy technology or country. This makes it difficult to compare data on greenhouse gas balances and cost-effectiveness of different biomass energy technologies. In Australia, a major biofuels study was undertaken (Beer *et al.* 2002) that established a standard framework for comparison of reference biofuels, although it did not lead to a software tool. One software tool developed in Europe enables user groups to compute and compare carbon benefits of biomass energy technologies throughout the EU. This was developed through the EU-funded BIOMITRE project. This study identified about 900 potential technology combinations for biomass energy projects in the European Union and established a standard method for conducting both biomass and reference system calculations (Horne 2005). Similarly the Greenhouse gases, Regulated Emissions, and Energy use in Transportation (GREET) model in the United States of America (USA) provides automated life cycle calculations for biomass and other fuel options (Argonne 2008).

However, tools have not delivered a high uptake, probably due in part to the complexities and resultant variations in goal, scope and system boundaries. One review study of biomass energy technologies for ethanol production found that, of two detailed LCA studies, one is generally unfavourable, while the other is significantly favourable (Blottnitz and Curran 2007). This study drew on 47 published LCAs that compared bio-ethanol systems with conventional fuels, and it concluded that choices about process residue handling and fuel combustion are important in explaining differences in results. These LCAs typically indicated that reductions in resource use and global warming can be achieved, although impacts on acidification, human toxicity and ecological toxicity, occurring mainly during the growing and processing of biomass, were more often unfavourable than favourable. This indicates one of the dangers of carbon assessment – that conducting carbon-only assessment involves the risk that technologies with other significant environmental impacts may be chosen.

10.3.3 Policy implications

Clearly, there is an extensive range of assumptions, uncertainties and situations in which biomass may be used for apparent carbon benefits. Many technologies and practices are untried, economically marginal or submarginal, or not being adopted optimally for carbon savings due to perverse incentives in the market. Given this situation, the current growth of interest and commencement of carbon trading schemes and various support mechanisms is appropriate in general terms. However, new policies and mechanisms must be carefully developed and based on 'complete' LCA-based information, taking into account all direct and indirect emissions and impacts, otherwise new perverse incentives may be created.

For example, a proposal for a biomass energy project was designed to generate electricity, displacing emissions from a coal-fired plant, and to use the sawdust and offcuts from a sawmill as feedstock (A. Cowie, pers. comm., 27 November 2007). The sawmill residues were being used as feedstock at a particleboard plant and, as part of the project, they were to be diverted to the biomass energy project. As a result, the particleboard manufacturer was required to use freshly harvested low-grade timber as feedstock, with particular implications for the particleboard plant. Because the green feedstock has a high moisture content and will need to be dried before use in particleboard manufacture, the dryer energy (usually supplied as natural gas) must be accounted for in a full LCA.

Calculations (Cowie and Gardner 2007) indicate that the increase in emissions from particleboard manufacture, due to the higher fossil fuel requirements to harvest, transport, chip and dry the green biomass, can amount to nearly 20% of the emissions displaced by the biomass energy project. The net emissions reduction would be 18% greater if the freshly harvested timber was used by the biomass energy project and the dry sawmill residues continued to be used for particleboard. This result is specific to the particular project in question. It is affected particularly by the fossil fuel displaced, the efficiency of the biomass energy project compared with the displaced coal-fired electricity plant, the moisture content of the sawmill residues, and the efficiency of the dryer in the particleboard plant. Whether the newly harvested timber is used for biomass energy or for particleboard, the carbon stock in the forest will be reduced, causing an additional indirect impact that further diminishes the net greenhouse benefit of the biomass energy project.

This example illustrates that indirect consequences can reduce the net greenhouse gas mitigation benefits of a biomass energy project. Such 'leakage' must be taken into consideration in devising incentives for renewable energy projects and in determining the credit earned by such projects.

In summary, there are various assumptions to be made in calculating the net savings in emissions associated with the substitution of fossil fuels by biomass. Each assumption, however, can greatly vary the overall 'saving' achieved. As an example, while ignoring the credit due to the avoided use of fossil fuel, one study into emission factors for electricity generation from burning wood produced a range from –1.9 kilograms of carbon dioxide per kilowatt hour (kg CO_2/kWh) to 6.8 kg CO_2/kWh (Reijnders and Huijbregts 2003). While the range of values was primarily a result of different assumptions about the extent of carbon sequestration, this nevertheless shows the potential magnitude of uncertainty associated with the calculation of emissions for substitution. Even when reporting for compliance with international standards, uncertainties remain in particular cases of carbon offset and biomass energy technology projects. Although these uncertainties do not overall constitute a case for pausing in the rapid adoption of such technologies and supporting policies, regulations and support mechanisms to tackle climate change, they do highlight the need for renewed and intense research and documentation of LCA studies relating to carbon sequestration. Even though the complexity of net carbon assessment for biomass projects is challenging for policy makers, studies have directly and usefully informed policy (e.g. as reported in Mortimer 2006), and policy must continue to be developed and implemented under conditions of 'best-estimate uncertainty' if climate change is to be tackled in a timely manner.

10.4   LCA and carbon management: a case study in industry change

Carbon assessment and greenhouse-neutral policies are now standard business practice for a growing number of large global companies. Orica is an example of an Australian-owned, publicly listed global company that is committed to becoming greenhouse-neutral, along with three other goals: zero-waste; 'environmentally friendly' operations, products and services; and being water-neutral (Orica 2007). This case study charts the role of LCA in assisting Orica to reach this point (after James *et al.* 2005).

Orica has four main businesses, all generally viewed as polluting: Mining Services, Fertilisers (Incitec Pivot), Chemicals, and Orica Consumer Products. Hence, the challenge for achieving a sustainable development path is significant. Two principal tools have been used by Orica in facilitating this challenge: LCA and The Natural Step. In 1990, Orica commenced its 'Challenge' program by creating milestones to reach and preferably pass in several key safety, health and environment areas within five years. The program has been renewed every five years with subsequent Challenge programs. In 2000, Orica Consumer Products undertook to conduct LCAs on all major product groups in response to the Challenge program. Detailed assessment work was undertaken on some paint products. This work developed in two principal ways: training in LCA of in-house staff, and commissioning of several LCA studies by the Centre for Design at RMIT University. The studies focused on:

- solvent-based and water-based paints (a core business in the decorative business unit)
- powder coating (a core business in Dulux Powders' business unit)
- No More Gaps (a core business in Selley's business unit)
- tinplate and plastic paint can packaging.

Each study was designed to inform:

- understanding of the environmental impacts of products and associated packaging
- environmental comparisons of existing products
- a selection process for raw material changes and future research projects.

A three-stage LCA strategy was adopted as follows:

- a preliminary LCA (on the nominated product) that utilised public LCA inventory data
- identification of important issues and areas where more localised data was necessary to obtain a more specific assessment
- data collection in-house and from suppliers and third parties to underpin more detailed assessments.

As a result, Orica has achieved both increased understanding of the environmental impacts of its products, packaging and processes and a positive environmental and capacity-building effect on suppliers. For example, Orica approaches led to Millennium Chemicals initiating its own LCA to assist it in understanding its product and production process and working with

Orica Consumer Products to reduce impacts throughout the supply chain. Orica Consumer Products has also been able to use the findings and data from the LCA reports to supply information to customers in tender submissions (e.g. green office fit-out questionnaires), providing a competitive edge. Along with these company benefits, there have been challenges in developing and accessing LCA expertise and in collecting appropriate inventory data in a timely manner.

In parallel to the LCA work, Orica Consumer Products considered the broader strategic context of sustainable development. While LCA helped quantify environmental impacts from product source to disposal ('cradle-to-grave'), it did not provide enough information on incidental impacts or impacts after use or disposal. In this regard, The Natural Step organisation was identified and employed as a vehicle for informing strategic decisions about the types of products to make, where new research needed to be directed and even what types of businesses to be in, given 'triple bottom line' concerns. Important questions included:

- what are the impacts of volatile organic compounds?
- what are the impacts of titanium dioxide use?
- what are the impacts of paint packaging waste generated at the consumer stage?

Senior managers from Orica Consumer Products business units participated in a series of workshops where findings from all four LCA studies were used to inform the testing of concepts and practicalities. Some ten initiatives emerged to further investigate the economic viability of sustainability-driven change, including:

- a project to improve efficiencies in the supply chain of titanium dioxide
- a 'Paintback' trial to reclaim and remanufacture post-consumer paint products, which led to a prestigious environmental award for the company
- a 'Green Office' waste and energy reduction initiative
- a car fleet fuel efficiency study
- a waste reduction search across the supply chain.

The benefits of using LCA at Orica are illustrated by the company's ongoing use of the technique. The current 2010 Challenge states:

> Greater focus will be placed on understanding the full life cycle impacts of a wider range of products. Wider use of [LCAs] will help show where reductions in waste, water and energy use may be made across an entire value chain (Orica 2007).

Moreover, through sustained use of LCA, supplemented with The Natural Step, Orica Consumer Products has developed confidence associated with increased knowledge of its practices. In turn, this has created a catalyst for innovation and sustainable value creation and informed a commitment to carbon neutrality.

10.5   Carbon accounting, LCA and future prospects

The difficulties and uncertainties in using bio-sequestration as a long-term strategy to address climate change are many and varied. Methodological issues remain in the literature, relating especially to the selection of allocation procedure and reference study parameters in calculating net carbon benefits. Nevertheless, the surge in interest and apparent commitment to greenhouse-neutral futures in recent times ensures that continued focus on valid methods for assessing appropriate technologies, substitutions, assumptions and calculation approaches will occur, leading to further improvements and consistency in results.

That several greenhouse gas accounting standards have been developed, implemented and approximately aligned, has provided confidence in comparative assessment, and smoothed the path for transactions-based greenhouse abatement, such as carbon trading, carbon taxation and carbon offsetting. These standards provide one practical approach to overcome difficulties and uncertainties in data and methods – an approach which contrasts with LCA practice. Indeed, the mere existence of carbon accounting standards could be seen as a failure of the LCA standards and LCA methods, since these existed previously, and could have been used more overtly as the basis for carbon accounting standards and practice. However, LCA standards leave open a range of methodological issues and approaches that create uncertainty and variable results.

This is not to say that LCA practice does not deal with uncertainty and methodological variations. On the contrary, it deals with these increasingly successfully through an overriding emphasis on openness and transparency, and rigorous sensitivity testing of the potential range of results arising from uncertain data or varied methodological approaches. This approach is essential for open, critical debate and development, for example, around allocation and boundary setting, which are discussed in standards but for which no definitive practice is prescribed. In contrast, the carbon accounting standards provide more specific rules, which truncate debate or methodological variations and thus gloss over many of the more difficult issues. The benefit is increased clarity and replicability. Moreover, these standards do cover many non-LCA related issues that could warrant standardisation.

Perhaps the larger failure of both LCA and carbon accounting standards is the lack of cross referencing of key methodological issues that are common to both. The magnitude of this failure is still to be determined and can be minimised if more attention is placed on methodological alignment through a 'best-of-both-worlds' approach. LCA's systemic approach contributes to the rigour required to underpin existing and future development of carbon accounting methods. Specific contributions of LCA could include boundary issues, in particular, the task of appropriate allocation of carbon burdens in complex systems and supply chains. This could lead to a functional, cumulative supply chain approach to emissions accounting, similar to the GST system.

Meanwhile, the clock is ticking and the window for early climate change action is closing – raising the pressure for a relentless focus on carbon. Carbon assessment and emissions reduction should not be pursued in a 'vacuum' to the detriment of other environmental issues, since this may lead to new perverse incentives and environmentally poor outcomes. Again, the growing history and practice of LCA provides a significant and useful body of knowledge and learning about how to assess and quantify environmental impact more holistically. LCA practice is familiar with problems of 'forest biodiversity versus embodied carbon savings of timber use' and can provide methodological consideration of the potential for burden-shifting (or problem-shifting) over time or between impacts from different abatement or efficiency projects. It can also bring the technique of consequential analysis to determine the additional effects of particular actions.

As the limits of carbon assessment are probed and revealed, so the prospects for LCA contributions to solutions will be raised. One possibility is that a reflective learning period will ensue, where evidence of burden shifting from greenhouse emission programs to other impacts will arise, and then more LCA-like methods will be used to improve the decision and accounting tools. There is already evidence of this in the biofuels debate, but it is as yet relatively absent from other carbon abatement forums. Beyond this, it is possible that widespread use of carbon offsets in affluent countries may cause a 'rebound effect' in people's behaviour. A sense that offsets 'solve' or 'atone' for greenhouse 'sins' may lead to carbon-intensive behaviour being allowed or even positively encouraged. Instead of looking for ways to reduce greenhouse gas emissions, consumers may use sequestration as a justification to continue existing practices unchecked, and emissions may increase as a result. Given the uncertainties and potential unreliability of sequestration over the long term, this would raise overall greenhouse gas levels compared to a strategy of reducing emissions at the source. This sobering reflection on sequestration is not simply a caution on the use of offsets, but also shows how important social factors are in any technical system – and underlines the fact that a more sophisticated approach to assessment and design of strategies for carbon abatement is urgently required.

In parallel to further methodological development is the need for rapid development of policy, regulatory and support mechanisms to provide confidence and viability to various industries to reduce greenhouse gas emissions. While the cost of carbon is already in common parlance, it will become more ubiquitous over the next decade and will have significant impacts on business and practice. Experience with data provision and LCA-related assessment, as indicated in the Orica case study, indicates that experience with LCA provides a good basis for organisations embarking on best practice carbon-reduction programs. It follows, therefore, that organisations that have begun to embed LCA as part of business management and decision-making will be well placed to take advantage of carbon savings in carbon trading or taxation environments.

10.6 References

AGO (2006) Greenhouse Friendly™ Guidelines. August. Department of Environment and Heritage Australian Greenhouse Office, Canberra.

Argonne (2008) *The Greenhouse Gases, Regulated Emissions, and Energy Use in Transportation (GREET) Model*. Argonne National Laboratory. Retrieved 14 April 2008 from <http://www.transportation.anl.gov/software/GREET/>.

Beer T, Grant T, Morgan G, Lapszewicz J, Anyon P, Edwards J, Nelson P, Watson H and Williams D (2002) 'Comparison of transport fuels: life-cycle emissions analysis of alternative fuels for heavy vehicles.' CSIRO, Aspendale, Victoria.

Beer T, Meyer M, Grant T, Russell K, Kirkby C, Chen D, Edis R, Lawson S, Weeks I, Galbally I, Fattore A, Smith D, Li Y, Wang G, Park KD, Turner D and Thacker J (2005) 'Life-cycle assessment of greenhouse gas emissions from agriculture in relation to marketing and regional development – irrigated maize: from maize field to grocery store.' Final Report HQ06A/6/F3.5, CSIRO Division of Marine and Atmospheric Research, Aspendale, Victoria.

Blottnitz HV and Curran MA (2007) A review of assessments conducted on bio-ethanol as a transportation fuel from a net energy, greenhouse gas, and environmental life cycle perspective. *Journal of Cleaner Production* 15(7), 607–619.

Council of Australian Governments (2005) Intergovernmental Agreement on a National Water Initiative. Council of Australian Governments, Canberra.

Cowie AL and Gardner WD (2007) Competition for the biomass resource: greenhouse impacts and implications for renewable energy incentive schemes. *Biomass and Bioenergy* 31, 601–607.

Department of Climate Change (2008) *National Greenhouse Accounts (NGA) Factors.* Commonwealth of Australia, Canberra.

Fraanje P and Lafleur M (1994) Verantwoord gebruik van hout in Nederland. IVAM Environmental Research, Amsterdam.

Gedalof Z, Peterson DL and Mantua NJ (2005). Atmospheric, climatic, and ecological controls on extreme wildfire years in the northwestern United States. *Ecological Applications* 15(1), 154–174.

Horne RE (2005) A new decision support tool for biomass energy technology projects in Europe. In: *Proceedings of the 4th Australian LCA Conference*, Sydney. Australian Life Cycle Assessment Society, Melbourne.

Horne RE and Matthews R (2004) *BIOMITRE Technical Manual.* Downloadable at <http://www.joanneum.at/biomitre/softwaretool>.

IPCC (2007a) 'Climate change 2007: the physical science basis.' Contribution of Working Group I to the Fourth Assessment Report of the IPCC. Cambridge University Press, Cambridge.

IPCC (2007b) 'Climate change 2007: impacts, adaptation and vulnerability.' Contribution of Working Group II to the Fourth Assessment Report of the IPCC. Cambridge University Press, Cambridge.

IPCC (2007c) 'Climate change 2007: mitigation of climate change.' Contribution of Working Group III to the Fourth Assessment Report of the IPCC. Cambridge University Press, Cambridge.

ISO (2006a) AS ISO 14064.1-2006 'Greenhouse gases – specification with guidance at the organization level for quantification and reporting of greenhouse gas emissions and removals.' Standards Australia, Sydney.

ISO (2006b) AS ISO 14064.2-2006 'Greenhouse gases – specification with guidance at the project level for quantification and reporting of greenhouse gas emission reductions and removal enhancements' (ISO 14062-2:2006, MOD). Standards Australia, Sydney.

ISO (2006c) International Organization for Standardization. AS ISO 14064.3-2006 Greenhouse gases – specification with guidance for the validation and verification of greenhouse gas assertions. Standards Australia, Sydney.

James K, Vockler R and Vandestadt S (2005) The Orica consumer products story of using LCA and The Natural Step. In: *Proceedings of the 4th Australian LCA Conference,* Sydney, February. Australian Life Cycle Assessment Society, Melbourne.

Mann MK and Spath PL (2001) A life cycle assessment of biomass cofiring in a coal-fired plant. *Cleaner Production Processes* **3**, 81–91.

Marland G and Schlamadinger B (1997) Forests for carbon sequestration or fossil fuel substitution? A sensitivity analysis. *Biomass and Bioenergy* **13**(6), 389–397.

Mortimer N (2006) The role of LCA in policy and commercial development: the case of liquid biofuels in the UK. In: *Proceedings of the 5th Australian Life Cycle Assessment Conference,* Melbourne. Australian Life Cycle Assessment Society (ALCAS), Melbourne.

Ney RA and Schnoor JL (2002) Incremental life cycle analysis: using uncertainty analysis to frame greenhouse gas balances from bioenergy systems for emission trading. *Biomass and Bioenergy* **22**(4), 257–269.

Oneil E, Lippke B and Mason L (2007) *Eastside Climate Change, Forest Health, Fire and Carbon Accounting.* Future of Washington's Forest and Forest Industries Study, Washington.

Orica (2007) 'Sustainability Report 2007.' Orica, Melbourne.

Reijnders L and Huijbregts MAJ (2003) Choices in calculating life cycle emissions of carbon containing gases associated with forest derived biofuels. *Journal of Cleaner Production* **11**(5), 527–532.

Schlamadinger B, Apps M, Bohlin F, Gustavsson L, Jungmeier G, Marland G, Pingoud K and Savolainen I (1997) Towards a standard methodology for greenhouse gas balances of bioenergy systems in comparison with fossil energy systems. *Biomass and Bioenergy* **13**(6), 359–375.

Stern N (2006) *Stern Review: The Economics of Climate Change.* Cambridge University Press, London.

van Dam J, Faaij A, Daugherty E, Gustavvson L, Elsayed MA, Horne RE, Matthews R, Mortimer ND, Schlamadinger B, Soimakallio S and Vikman P (2004) Development of standard tool for evaluating greenhouse gas balances and cost-effectiveness of biomass energy technologies. *2nd World Conference on Biomass for Energy, Industry and Climate Protection in Rome, 10–14 June.* CD-ROM.

Westerling AL, Hidalgo HG, Cayan DR and Swetnam TW (2006) Warming and earlier spring increase western U.S. forest wildfire activity. *Science* **313**(5789), 940–943.

WRI-WBCSD (2004) *The Greenhouse Gas Protocol: A Corporate Accounting and Reporting Standard.* Revised Ed. World Resources Institute, Washington DC.

WRI-WBCSD (2005) *The GHG Protocol for Project Accounting.* World Resources Institute, Washington DC.

Chapter 11

Accelerating life cycle assessment uptake: life cycle management and 'quick' LCA tools

Ralph E Horne and Karli L Verghese

11.1   Introduction

Although the traditional life cycle assessment (LCA) study provides detailed assessment of the system under study, typical limitations include long lead times in data collection and analysis. There is then the question of how to translate the outcomes from a technical LCA report into decisions and changes in business systems. Easier and quicker ways of doing such assessments and producing 'decision-ready' LCA outputs may therefore accelerate LCA uptake under certain circumstances. In turn, these may provide a 'way in' to enable life cycle thinking and LCA results to be embedded within commercial systems. Life cycle management (LCM) systems and 'quick' LCA tools are examples of potential 'acceleration' aids. Of course, the effectiveness of these tools will be determined by the accuracy needs and possibilities, and the particular decisions to be supported. They are likely to be designed for specific industry sectors to enable relevant support needs to be met. The ultimate usefulness of these tools will be determined by a set of requirements including accuracy, functionality, reliability, validity and useability (Verghese *et al.* 2009).

In this chapter, LCM and 'quick' LCA tools are examined as candidates for aiding the acceleration of LCA uptake, primarily in commercial settings. This examination includes a review of the needs and roles of key stakeholder groups in affecting LCA uptake, and from this, a set of design requirements for 'quick' LCA tools. Two case studies of such tools are developed to illustrate both their form and the importance of context and stakeholders in tool development (see Sections 11.4.1.and 11.4.2).

11.2   Life cycle management overview

While the focus of LCA over the past four decades has predominately been refining methodology, data collection and data quality, there has been increasing interest more recently in LCM and its 'broader approach and focus on the application of and education on LCA and life cycle thinking' (Heinrich and Klopffer 2002, p. 315). According to the International Life Cycle Initiative, LCM:

> is not a single tool or methodology but a management system collecting,
> structuring and disseminating product-related information from various
> programs, concepts and tools (Remmen *et al.* 2007, p. 5).

The need to define and incorporate LCM into business and government decision-making is multi-fold. First, individual LCA studies are time and resource intensive. Second, they are invariably designed around a specific functional unit and particular questions that are temporally and spatially dependent. Third, they do not automatically lead to the embedding of life cycle thinking into the operations of business and governments. The latter is essential if we are to ensure that a full life cycle perspective is taken into account when designing and producing products or policies which comply with sustainable development principles. LCM can be viewed as a means by which the economic, technological and social aspects of products are integrated and continuously optimised (Weidema n.d.). Improved product quality, new technological innovations, opportunities to reduce costs, new or enhanced laws and regulations or pressure from customers and suppliers for products and services with improved environmental performance are examples of factors by which an organisation may decide to integrate LCM within their core activities. Furthermore, the successful implementation of LCM will require agreement on corporate policies and strategies to reduce impacts, the development and use of environmental assessment tools to guide decision-making, the integration of eco-design into product development processes, the implementation of supply chain management, and environmental communications and reporting of processes, products and deliverables (James 2004).

The essential logic of LCM is that if LCAs could be undertaken quickly, simply and reliably, and the information systematically collated and used in an active management system, then LCA information could be more widely used in decision-making, leading to improved environmental performance in a commercially efficient way. Interest in LCM tools has risen as the need for easily accessible information on embodied energy of materials, greenhouse gas emissions from processes, waste and recycling statistics and material selection continues to increase in sectors such as the packaging industry, building and construction and furniture and fit-outs (Verghese and Hes 2007). Software tools and calculators that provide these services have proliferated to fill this need. These 'quick' LCA tools vary widely in size, shape, look and feel, and areas of application, while they share core data and algorithmic combinations with LCA methods and techniques. Ranging from simple spreadsheets to online software programs with slick interfaces, they are increasingly used by engineers, environmental managers, designers, non-specialists and lay people.

One specific and important role for 'quick' LCA tools lies with their potential to facilitate widespread use of LCA information through the design process. The history of the design and development of such tools is relatively recent and extends back to the application of ecologically sustainable design ('eco-design') principles in the 1980s. In order to optimise the use of such principles within daily decision-making, some form of quick impact calculations were needed. As an example, William McDonough collaborated with office furniture manufacturer Herman Miller in the late 1990s to create one of the first practically applied product assessment tools, in that case to evaluate progress towards cradle-to-cradle products (Rossi *et al.* 2006). Figure 11.1 illustrates how different departments within an organisation can contribute to an LCM program.

One of the world's leading aluminium companies, Alcan, uses LCM for various applications including benchmarking aspects of environmental performance of their products with competitors, improving their internal and supply chain environmental performance, providing marketing information, and for strategic planning (Rebitzer and Buxmann 2005). Rebitzer and Buxmann (2005) state that five indicators are typically reported from Alcan LCA studies: primary energy demand, global warming potential, eco-indicator score (without energy and global warming), waste generation, and water consumption. These have been selected as the most relevant to the company, while being easily communicated to other parties and decision-makers within the organisation.

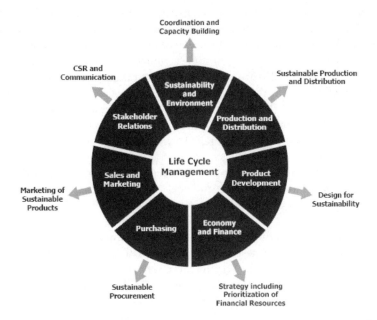

Figure 11.1   All functions have an important role in life cycle management (Remmen *et al.* 2007).
CSR, corporate social responsibility.

11.3   Stakeholders and the development of 'quick' LCA tools

Some 'quick' LCA tools have enjoyed popularity as online check points for greenhouse gas emissions, attracting thousands of visitors, while others have been the subject of considerable investment in development but with limited uptake upon completion. In any case, the design, production, marketing, consumption and end-of-life waste management of modern products and services typically involves dozens of stakeholders, each with different interests and influences on the system and its outcomes. From financiers, marketers, brand-owners and retailers to manufacturers and designers, consumers and consumer organisations, public policy makers and regulators, there is a considerable and complex set of interrelationships in any given product provision or consumption system. These groups, organisations and individuals are important not just as a 'context' for 'quick' LCA tools, but as key actors determining the system itself, and therefore determining the environmental impact of the resulting products and services. The key stakeholders are grouped as:

- designers and manufacturers
- consumers, purchasers and specifiers
- policy makers and regulators.

11.3.1   Designers and manufacturers

Designers often report that they would like to incorporate LCA into their design processes. This sometimes proves difficult or simply does not eventuate due to a lack of time to obtain the information (or an apparent lack of information); a lack of training or capacity to source and use LCA information; and/or a lack of interest from clients (Lewis and Gertsakis 2001). Clearly, there is a *prima facie* case here for 'quick' LCA tools, since they could potentially facilitate the provision of LCA information in a timely manner. A lack of interest from clients may be related to lack of markets or a perceived lack of market drivers, with inertia having a role in the latter.

We will return to this issue below. Meanwhile, industrial design professions are not well equipped with in-service training, and until 2007 there were no nationally available materials in Australia to assist tertiary programs to incorporate eco-design into curricula (Centre for Design 2007). Despite a decade or more of some form of LCA information being made available to designers, the piecemeal nature of this information and lack of any clear and simple means to access and use it effectively add to the reasons for low uptake of life cycle thinking into design processes. Due to the nature of LCA as it developed in the 1980s and 1990s, with shortages of data and controversy over allocation rules and other methodological issues, the technique gathered a reputation for being relatively technical, expensive and limited. In many cases, a comparison of just two product scenarios was realistic and still relatively time-consuming. From anecdotal evidence gathered by the authors, it would appear that many firms involved in design and/or manufacturing still retain this image of LCA, despite its rapid development. Accordingly, the incorporation of LCA is often considered to be a relatively unimportant niche activity.

Despite the difficulties and ill-defined market drivers, some manufacturers have embraced LCA across a range of sectors. These include commercial furniture, mass-market electrical consumer goods, the automotive industry and even service sectors (Gradael 1997). The development of LCA and cleaner production techniques and processes has led to a range of resource efficiencies in manufacturing processes, from raw materials production to fabrication of complex consumer goods. However, in this process it is necessary to differentiate between the use of environmental thinking during design, as in eco-design, and the (comparative) assessment of environmental performance after design, as in traditional LCA. Where early intervention in the conceptual design phase can be achieved, there is maximum opportunity for impact reduction by incorporating product life cycle thinking (Lewis and Gertsakis 2001), and it is here that potential exists for quick LCA tools to influence product design.

11.3.2   Consumers, purchasers and specifiers

Consumers exhibit complex behaviours. Many people view themselves as 'green' consumers, but relatively few act consistently green, as evidenced by the market share of environmentally preferable products and the observation that most voluntarily labelled products are not market leaders (OECD 2005; Pedersen and Neergaard 2006). In Australia, the 'success stories' in consumer choice of environmental products centre on clear, informative, independent information about products, and issues where there has been significant media focus. For example, Australia's energy labelling program for white goods is widely regarded as among the most informative and successful in the world; 94% of consumers recognise it and 88% use it in purchasing (Artcraft Research 2005). The main motivation for this is mixed. Helping the environment (13%) ranks below cost savings (39%) and energy savings (38%) in this regard.

Although cost and price are important issues for consumers, they are not the only ones. Purchases are often guided by quality or habit (Gallastegui 2002). Other key factors include (Hemmelskamp and Brockman 1997):

- consumer satisfaction
- values
- identification
- cost
- availability
- social pressure and consumer boycotts.

Green purchasing is a complex process, given the dynamic context of diverse purchasing situations (Manzini *et al.* 2006). There may be lack of information, and consumers may not understand how to purchase appropriately for a specific environmental problem (Szarka 1991). Moreover, environmental consciousness does not necessarily lead to environmentally friendly behaviour, and environmental awareness does not necessarily lead to a change in purchasing behaviour (Gallaroti 1995; Pedersen and Neergaard 2006). Different types of green consumers exist; some may be 'selectively green' and/or may be manipulated to purchase products that are not green because of imperfect information (Pedersen and Neergaard 2006). What is perhaps most significant is that information overload is clearly rife in consumer purchasing. In one study, 97% of those surveyed indicated that there 'was more stuff to read than I could ever dream of reading', and 92% indicated that they felt 'surrounded' by information (Lloyd 2006). Of particular interest for quick LCA is the issue of availability and trustworthiness of environmental information. While there is good recognition of eco-label-type information in the Nordic countries (Leire and Thidell 2005), there is poor recognition and a lack of understanding of symbols and labels in Australia (e.g. in New South Wales (Taverner Research Company 2004)). The exceptions are the Minimum Energy Performance Standards energy labelling and Water Efficiency Labelling Scheme, which are both mandatory and both enjoy widespread recognition among consumers (Horne *et al.* 2007).

Key barriers for consumers include price, awareness, trust, the complexity and availability of information, and the interplay of this with other behavioural factors and influences. These issues also apply to business-to-business 'sustainable' purchasing, in which there is evidence of an increase in activity (AELA 2004). Indeed, on the international front, eco-labelling programs in New Zealand and Canada have turned their focus from the consumer market towards the professional purchasing market. For both business and household consumers, there is a case for trustworthy, clear, useful and independent environmental information, and quick LCA tools could be involved in the supply of such information, perhaps incorporating some form of labelling.

Specifiers are a particular group involved in purchasing. For example, interior designers are often responsible for specifying purchases they will not themselves make, but for which they exercise professional responsibility on behalf of clients. In common with other 'professional' purchasers, their decisions are generally more informed and more prone to influence by business drivers. In Australia, as in other western countries, specifiers are engaged in responding to an emerging set of 'green building' drivers (which vary in their relationship to, and their use of, LCA information). Since specification goes beyond purchasing and involves considerable influence over the nature of the products purchased, there is a specific role for quick LCA tools in assisting specifiers. For example, in determining whether to specify wool or synthetic textiles for chair upholstery, such a tool may allow a specifier to determine reliably and quickly the best environmental option – and the environmental significance of any particular decision.

11.3.3 Policy makers and regulators

As indicated above, price is an important issue for consumers. Regulation helps to determine prices through a range of market-setting, calibration and intervention strategies. On this basis, if regulation required the external environmental costs of all processes to be incorporated into prices by adding the value of the damage they cause, then many environmentally benign goods and services would be cheaper than they are at present. While a summary of the wide and contested literature of neo-classical environmental economics is outside the scope of this book, the point is that regulators have a major role in determining the unevenness of playing fields for environmentally preferable goods and services.

Since governments have not generally implemented wide-scale market mechanisms to internalise environmental externalities in consumer goods, there is sometimes a price premium on environmental products. This is despite them having a lower cost to society (including labour, material, social and environmental costs). This amounts to ambivalence by policy makers and governments. Given this situation, it is perhaps unsurprising that research suggests consumers are not convinced of the importance of their contribution, and mainly expect legislation to be set by public authorities first (Zaccaï 2008). This may imply an aversion to being in the minority of 'payers' (environmental product purchasers) while the majority remain as free riders who get cheap goods, and meanwhile everyone pays the consequences through climate change and other impacts.

Short term economic and political perspectives also give businesses a false indication that long-term environmental risk assessment, LCA, product stewardship and corporate social responsibility are not important. The net result is that a lack of good regulation to guide purchasers and businesses in responding to environmental concerns is a significant factor in preventing demand for environmental purchasing, and also presents a barrier to the uptake of quick LCA in informing both design and purchasing to this end.

11.4   Design requirements for quick LCA tools

From the preceding analysis of stakeholder needs regarding drivers for eco-design, provision of environmental information and uptake of environmental product purchasing, there is a potential role for quick LCA tools, provided they meet one or more of the following conditions; that they:

- assist designers and manufacturers to make early design and manufacturing decisions based on LCA information
- assist environmental managers to control the environmental impacts of organisations by providing quick LCA information to help them meet environmental objectives and targets
- provide information in clear, simple, accessible ways for use by consumers, purchasers or specifiers
- engender or contribute to a system of trust and independence in the supply of information about environmental performance
- provide businesses with the means to comply with regulations and codes of environmental performance
- provide policy makers with clear LCA information to inform policy and regulatory development.

In meeting such requirements, however, any quick LCA tool must address competing needs for: (a) sophistication in consideration and evaluation of environmental issues, and (b) simplicity, ease and speed-of-use by different actors in design, production and/or consumption. Compromise is inevitable with any quick LCA tool. Careful validation is required to provide confidence that LCA results can be replicated with automated quick LCA kits to acceptable levels of accuracy before they will be widely adopted as trusted tools. These competing needs partly explain the wide range of approaches, from simple principles-based 'checklists' to spreadsheet or online algorithm-based calculation software programs that provide automated reports. The latter hold out the possibility of more accuracy and ready application, yet may not be transparent enough and are often only applicable to a particular sector or relatively narrow range of design situations.

The development and use of quick LCA tools would fast-track the integration of environmental design aspects into the product development process (Byggeth and Hochschorner 2005). The successful implementation and use of these tools will be guaranteed if they meet a range of requirements (Lofthouse 2006), that:

- they focus on design and the design process and not just on strategic management or retrospective analysis of existing products
- they combine guidance, information and education
- the content is both specific and broad with respect to, for example, the industry sector and materials
- they are streamlined and can be easily incorporated into daily practices
- information is presented in easy-to-digest pieces of information
- it compliments the way the user works and can be referred to as and when required.

The following sections investigate the development of two specific quick LCA tools: one for the packaging supply chain and the second for product designers.

11.5   Examples of LCA tools

As indicated in the introduction to this chapter, product and industrial designers have been using checklists and principles to guide environmental design considerations for several decades. There are some excellent, comprehensive manuals and handbooks to assist the design community (e.g. Yeang 2006). The uptake in eco-design remains patchy, but is gathering momentum in the face of customer demand. Indeed, in Australia and internationally:

> there is a growing demand for certification of Type I or Type III eco-labels from influential customers such as retailers, manufacturers and government purchasing programmes (D'Souza *et al.* 2006).

However, there is also a general and growing need for more detail and accuracy in the information. Not least, this is needed to deal with counter-intuitive results arising from misapplication or misunderstanding of impacts, such as the widely held view that 'natural' materials such as wool or leather must always be environmentally preferable to synthetic alternatives. The role for quick LCA tools in product design extends well beyond addressing such misapprehensions.

A product-design LCA tool could sit at the intersection between the initial conceptual design process, the regulatory environmental agenda and the provision of environmental performance information to consumers, purchasers and/or specifiers. Inevitably, a similar set of needs for quick, accurate, transparent and independent performance-based information exists, with additional requirements relating to the nature of the products to be evaluated.

11.5.1   Case study 1: The case for a packaging tool

Consumer packaging has been under scrutiny since the 1980s as a potent symbol of single-use conspicuous consumption culture. Nevertheless, packaging delivers many functions, from protection and containment to shelf appeal, and almost every product grown and manufactured in modern society is packaged at some stage in its life. Food and beverages are the major users of packaging, and as the availability of different foodstuffs increases, so does the quantity and complexity of the corresponding packaging materials around them.

Traditionally, used packaging across the western world has become waste, and this provides a significant material loss in the industrial system. Used packaging waste in the European Union (EU) alone will reach an estimated 77 million tonnes by 2008 – an increase of 18% from the waste generated in 2000 (Monkhouse *et al.* 2004, p. 13). Clearly, considerable attention is needed to reduce this material loss. In cash terms alone, world packaging value is estimated to be US$300 billion per year, with Australia's contribution valued at A$7 billion per year to 7.5 billion per year (Packaging Council of Australia 2005). As the world industrialises, so the problem of packaging material loss expands. In Asia, it is estimated that by 2025, 1.8 million tonnes of waste will be generated per day (Hoornweg 1999), with a shift in waste composition towards more paper products and a much higher proportion of plastics and multi-material items. Newspapers, magazines (along with corresponding increases in advertising), fast-service restaurants, single-serve beverages, packaged ready-foods and mass-produced products all mean more waste potential.

The environmental policy response to packaging originates from increasing concern over rising rates of litter, landfill and material losses from packaging waste. This has led to the development of regulations and waste policies by different governments. The underlining focus is a shift to 'product stewardship' – the sharing of responsibility to reduce impacts on the environment throughout the supply chain. In Australia, New Zealand and the United States of America (USA) the preferred model is 'shared responsibility' while it is more common in Europe to see 'extended producer responsibility' (i.e. producers finanically responsible for waste management). In the EU, there are binding regulations, such as the Directive on Packaging and Packaging Waste. Accompanying the directive are Essential Requirements such as waste prevention measures, possible packaging indicators, waste prevention plans, reuse, producer responsibility and issues related to heavy metal pollution. Other common policy tools include container deposits and eco-taxes on particular packaging materials. The policy trends in Asia, including Japan, China, Taiwan and South Korea, are recycling targets, take-back requirements and mandatory recycling fees or restrictions on particular materials (Lewis 2006).

In Australia, the National Packaging Covenant (NPC) is the voluntary component of a national co-regulatory approach between government and industry to the LCM of packaging throughout the supply chain. The scope includes impacts across all stages of production, distribution, use, collection, reuse, recycling, re-processing and disposal. The NPC aims to improve the total environmental performance of consumer packaging and distribution packaging, and is underpinned by a regulatory framework, the National Environment Protection Measure (NEPM) for Used Packaging Materials, which is implemented by each state or territory government in Australia. The NEPM is designed to deal with non-signatories and non-compliant signatories. Implemented in 1999, the NPC was reviewed in 2004, and a new covenant took effect in July 2005 that included key performance indicators (KPIs) and targets requiring business in all industry sectors to increase their commitment to reducing packaging's environmental impact, and to provide evidence of achievements and due diligence in changes to packaging system design. Examples of KPIs for companies include (NPCC 2005):

- changes in the total weight of consumer packaging and the total weight of products packaged in the Australian market
- changes in the average post-consumer recycled content in packaging
- changes in the total weight by type of 'non-recyclable' packaging in the Australian market
- the amount of consumer packaging in the total waste stream and its relativity to other waste stream components.

In addition, the Environmental Code of Practice for Packaging (ECoPP), which forms Schedule 5 of the NPC, has been updated to provide improved guidance on the role and importance of eco-design within the packaging supply chain.

Such regulations provide a key driver for the packaging supply and waste management chains – reducing materials used in packaging and reducing the amount of packaging material discarded in landfill. However, the climate change agenda is also now being added to the material recovery agenda, along with a range of other environmental issues that more traditionally lend themselves to evaluation using LCA. This provides an initial case for a quick LCA packaging tool. But what should this tool look like, what functions should it perform, and how?

Table 11.1   Selection of packaging decision-support tools (Verghese *et al.* 2006)

Name of tool	Features of tool
Packaging Impact Quick Evaluation Tool (PIQET)	Evaluates all stages of the life cycle for the complete packaging system and all its packaging components per pallet. Reports against key environmental indicators and packaging specific indicators (refer to Table 11.2). Delivered as a web-based tool.
Tool for environmental Optimisation of Packaging design (TOP)	Evaluates packaging in conjunction with the product in light of the essential requirements of the EU directive. Indicators considered: product-packaging combination, added value, logistics efficiency, heavy metals, reuse and recovery, material consumption, environmental impact. Delivered in a spreadsheet format.
MERGE™ (also known as COMPASS) tool by Environmental Defense (also referred to as The Sustainable Packaging Coalition) in the USA now has exclusive license for the packaging design aspect of this tool	Focuses on formulated goods such as personal care items and household/professional cleaning and maintenance products. Metrics for packaging include: resource consumption, energy consumption, virgin materials content, non-recyclable materials content, 'bad-actor' packaging, greenhouse gases and pallet inefficiency.
Wal-Mart Scorecard	Aim is to increase the percentage of packaging made from renewable resources by replacing non-recoverable materials. The tool calculates raw scores of packaging based on: packaging material, production, transportation. Other factors considered include: recycled content, renewable energy resources. Limitations are that the tool gives raw score, rank and weight, which are not readily transparent.

Following from the above discussion, any quick LCA tool should be able to differentiate between packaging options on the basis of key environmental impacts, such as materials and energy use and greenhouse gas emissions. It should be quick and accurate enough to be able to be used by packaging technologists, designers and manufacturers in making early design and manufacturing decisions. It should also provide information in clear, simple, accessible and independent ways, with the potential also to provide useful information for environmental managers, policy makers and consumers. Finally, it should provide businesses with the means to comply with regulations and codes of environmental performance (e.g. those relating to packaging and waste reduction). This latter point is important, since such regulations have developed in various countries in recent years, but companies still have limited access to information on the life cycle impacts of their packaging without a straightforward, efficient scientific way of strategically addressing packaging sustainability.

Not surprisingly, there have been a range of attempts to fill this gap. A range of decision-support tools have been developed to assist those in the packaging supply chain to consider the environmental impact of packaging options. A selection of tools produced worldwide by industry associations in conjunction with individual companies, government departments or by individual firms are presented in Table 11.1.

The Packaging Impact Quick Evaluation Tool (PIQET) project began in November 2004 with the aim of demonstrating through direct application to a sponsoring company's case studies that rapid and credible environmental assessment can be performed for food packaging systems within the Australian context (Horne *et al.* 2005; Verghese *et al.* 2006). The Sustainable Packaging Alliance (SPA) developed a stakeholder group of sponsors and advisors that included Sustainability Victoria, Cadbury Schweppes, Lion Nathan, Nestle Australia, MasterFoods

Australia and New Zealand, and Simplot Australia. With additional grant support from the Australian government through the Department of the Environment and Water Resources, and the Department of Communications, Information Technology and the Arts, a prototype tool was developed to confirm the algorithms, data requirements and test functionality. A full Internet-based tool was then commissioned and developed (programming undertaken by WSP Environmental), opening for subscriptions in September 2007 and publicly launched in March 2008. A program of further research and development is underway (2009) to roll out functionality globally and across other sectors.

Importantly, packaging technologists and environmental managers were consulted throughout PIQET's development, through regular meetings with the research team to identify the features and usability of the tool. The needs of other stakeholders and business drivers were also considered, along with the functional needs criteria discussed above. The resultant PIQET version 1.1 provides a 10 minute to 20 minute turn-around time and:

- assesses the environmental impact of different packaging formats
- evaluates and compares new and existing packaging systems and materials, and explores for better options
- identifies environmental issues associated with particular materials at an early development stage
- measures and reports against a range of environmental performance indicators for internal and external stakeholders such as customers and regulators
- benchmarks packaging performance over time
- sets targets, standards and specifications
- integrates environmental decision-making into the packaging design process
- identifies priority areas for improvement, for example:
  - the points where impacts occur in the life cycle
  - which packaging components or systems (e.g. the subretail, retail, merchandising and traded unit levels) have the highest impacts
- evaluates the effect of recyclability and recycling rates on the overall environmental impact of packaging materials
- provides due diligence in demonstrating a range of options in the design stage
- can be used to facilitate discussions with supply chain partners and other stakeholders for purposes of education, awareness building and negotiation.

Individual packaging components that make up the packaging system are modelled in PIQET at the subretail, retail, merchandising or traded-unit level (Fig. 11.2).

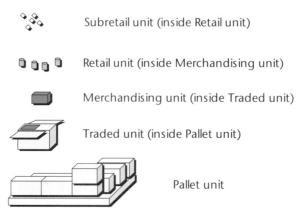

Subretail unit (inside Retail unit)

Retail unit (inside Merchandising unit)

Merchandising unit (inside Traded unit)

Traded unit (inside Pallet unit)

Pallet unit

Figure 11.2   Standardised packaging format levels used in the Packaging Impact Quick Evaluation Tool (PIQET).

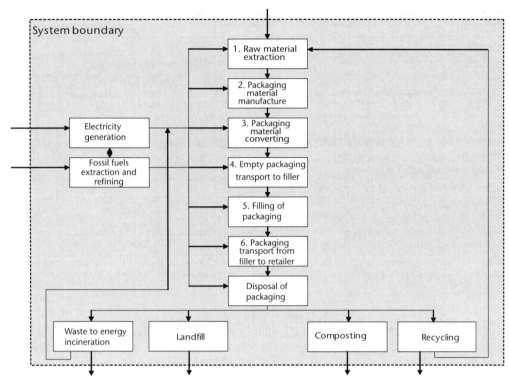

Figure 11.3   The Packaging Impact Quick Evaluation Tool (PIQET) system boundary.

The user enters data for the packaging life cycle stages presented in Figure 11.3. The need for the user to enter data is minimised through the optimisation of calculators and algorithms to convert mass to volume and so on, while drop-down menus facilitate selections and provide a quick, intuitive interface (Fig. 11.4).

Components in packaging system

Define all the packaging components used to package the product and indicate how each component relates to the packaging levels you defined above. Make sure you enter every component. If you need to add more components click the link below.

Tip: Enter each component from the pallet level down to the retail or sub-retail level.

No.	Type of component	Type of material	Level of packaging	Weight of one component
				gram
1	Bottle	Glass - green	Sub-Retail unit (bottle)	205
2	Closure	Metal - steel	Sub-Retail unit (bottle)	2.06
3	Carton	Board - cardboard corrugated	Merchandising unit (carton)	240
4	Mulitpack	Board - cardboard high wet stre	Retail unit (Sixpack)	35.6
5	Label	Resin - PP (film grade)	Sub-Retail unit (bottle)	1.2

Add component

Figure 11.4   Screen shot indicating selection of component type and material in the Packaging Impact Quick Evaluation Tool (PIQET).

Table 11.2   Life cycle environmental impact and packaging-specific indicators reported in the Packaging Impact Quick Evaluation Tool (PIQET)

Packaging specific indicators	
• Product/packaging ratio • Product % remaining in packaging • Packaging to landfill as a % and kg • Packaging to recycle as a % and kg • Recycled content % of packaging per pallet load • Packaging as a % of packaged product weight • Mass of packaging recyclable	• kg and % of packaging per packaging level (subretail, retail, merchandising, traded and pallet) • Recycled content of each individual packaging component • Packaging material summary (no. of each individual packaging material in packaging system format)
Life Cycle Assessment (LCA) environmental indicators	
• Global warming (kg CO_2 eq.) • Cumulative energy demand (MJ LHV) • Minerals and fuel (MJ surplus) • Photochemical oxidation (kg C_2H_4 eq.)	• Eutrophication (kg PO_4^{3-} eq.) • Land use (hectare) • Water use (kL H_2O) • Solid waste (kilogram)

kg C_2H_2 eq., kilograms of ethane equivalents; kg CO_2 eq., kilograms of carbon dioxide equivalent; kL H_2O, kilolitres of water; kg PO_4^{3-} eq., kilograms of phosphate equivalent; MJ LHV, megajoules of lower heating value; MJ surplus, megajoules of surplus energy.

The key results calculated by PIQET include a range of life-cycle environmental impact indicators and further packaging-specific indicators (Table 11.2). LCA impact categories provide a credible and rigorous environmental assessment, while other indicators allow the tool to report against existing requirements for the NPC and ECoPP.

For each packaging system format entered, the outputs are reported in both tabular and graphical formats. Figure 11.5 shows, as an example, the carbon dioxide emissions for a particular packaging format. For this case, the subretail packaging contributes the most to global warming, while the retail cluster and traded-unit packaging make relatively minor contributions. Moreover, for the subretail packaging, material production is the main contributor to

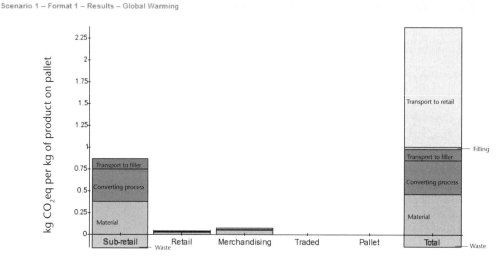

Figure 11.5   Global warming results graph for the different packaging component levels and life cycle stages. (All positive results are impacts and all negative results are benefits. Only the 'total' includes transport to retail and filling.)

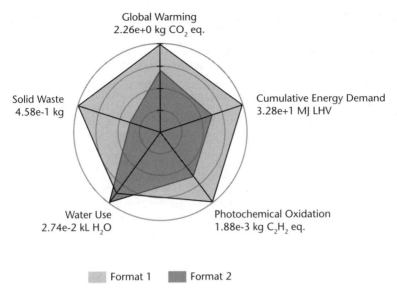

Figure 11.6   The relative impact of selected scenarios on equally weighted indicators. The closer to the centre, the lower the impact.

carbon dioxide generation, followed by the converting process. The negative result shown in the figure is an environmental benefit and is due to the benefits gained in recycling of the material for this particular format.

Figure 11.6 provides a comparison of two different packaging formats against five different impact parameters: global warming, solid waste, cumulative energy demand, water consumption and photochemical oxidation (smog). PIQET automatically normalises the results so that the axis for each parameter extends to the highest scored result. This allows users to make their own trade-offs and comparisons according to their own company policies and business drivers, rather than forcing a particular policy objective by prioritising one issue over another. With this in mind, the results show that, in this example, format 2 is the best performer across all five impact categories, except water use.

PIQET was developed to meet a range of functional requirements (Table 11.3), and these have been met or exceeded in the first version of the Internet-based tool.

11.5.2   Case Study 2: Development of a product-design quick LCA tool

Product eco-design strategies are increasingly drawing on both eco-design principles and LCA data. However, there is a particular need for quick LCA information due to the nature of typical product design processes. Concept design, which establishes key aspects of look, feel, functionality and aesthetic quality, effectively also 'locks in' many environmental burdens early in the design process. Therefore, waiting for detailed design before undertaking detailed LCA misses out key opportunities to 'design out' some environmental burdens.

The need for a concept-stage quick LCA product design assistant led to the research and development of Greenfly, a web-based software tool based on LCA algorithms and data, but with an intuitive, easy-to-use interface. Greenfly allows product designers to improve environmental performance of products, from design concept through materials selection to detailed design. As with PIQET, a collegiate approach has been adopted in the Greenfly development process, with the core research team led by the Centre for Design at RMIT University and including WSP Environmental, the Design Institute of Australia and Sustainability Victoria. A

Table 11.3 Functional requirements met by Packaging Impact Quick Evaluation Tool (PIQET)

Requirement	Features
Online access and use	Internet-based, security/protection of company information, ability to import archived files/data
Quick turn-around	No longer than 20 minutes per assessment, default data for non-available specific data
Easy to use	Limits information input by operator, menu-driven, 'Help' function
Quick scan versus detailed evaluation	Ability to short-cut and input specific data
Environmental focus	Integrates with company environmental management strategy (menu choice list to select company objectives/indicators), integrates with National Packaging Covenant KPI's, targets and reporting, demonstrates use of the Environmental Code of Practice for Packaging, benchmarking ability on total packaging system and individual components
Scientific credentials	Life cycle assessment-based, case study validation against life cycle assessment studies

KPI, key performance indicators

wider stakeholder group consists of commercial furniture manufacturers, industrial designers, specifiers, and government and non-government stakeholder organisations.

Greenfly Online was developed from a paper-based tool which adopts a five-step approach. The first step is the design concept, which involves high-level tracking of major material flows from origin to end-of-life/recycling. Visual techniques and recording of the design process are encouraged, along with conceptual comparisons. The second step involves the application of common rules to identify which phases of the life cycle have the greatest overall impact, along with relevant/potential environmental drivers for each phase, such as consumer demand, industry/client needs or regulations. The third step involves further life cycle environmental impact assessment at a more detailed level, reflecting the progression of the design process. While not discounting the use of 'full' LCA in this process, Greenfly Online also provides semi-quantitative estimation methods to follow as a means to achieve the assessment. Environmental-indicator scores and basic information for common materials and processes are included to assist in this. With an initial design and evaluation complete, the fourth step provides life-cycle eco-design strategies to enable the designer to re-assess, re-design and re-select materials to optimise environmental performance. Finally, the fifth step requires review and reflection, plus the opportunity to complete documentation of the process and outcome.

Greenfly Online uses web 2.0 development technology and requires minimal user input to translate LCA data into a selection of key environmental indicators displayed in real-time on screen and in report format. Individual components are presented in a 'tab' style format from general product information (e.g. name of designer, design brief and product images) to the product life cycle, with individual pages for each life cycle phase:

- manufacture – material selection and manufacturing stage (Fig. 11.7)
- transport – defining transport options
- use – defining operational aspects
- end of life – detailing what happens at waste management.

An innovative aspect of Greenfly Online is 'real time' results data, presented as graphs on the right-hand side of the page as the user develops a design scenario. As illustrated in Figure 11.7, a pie chart located on the right-hand side of the screen indicates the relative burdens of different life cycle phases. The user can select which environmental indicator they wish to view

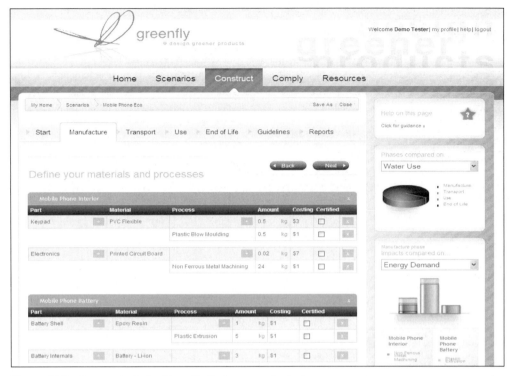

Figure 11.7 Screen shot of a Greenfly Online manufacturing page.

instantly at any time, from the following four: global warming, water use, energy use and solid waste. A bar chart is located immediately below the pie chart, and this compares impacts within the life cycle phase the user is considering. Both graphics refresh instantly each time a page is changed, a value is entered or a selection is made.

Enrichment of the data entry and calculations process is provided in several ways. The designer is presented with a range of design guidelines and prompt questions to be considered as they select materials and work through the life cycle stages (Fig. 11.8). Those fields that are completed are collated and appear in automatically generated reports. Guidelines that address eco-design considerations are also included, both as quick tips and as links to a catalogue of downloadable resources in portable document format (PDF).

A range of resources is available as downloadable PDF files – easily accessible for the user and also for uploading updates (Fig. 11.9).

The Greenfly Online structure also allows for the development of a set of modules for different design sectors. For example, a survey of drivers in relevant sectors revealed that commercial furniture is a suitable sector to develop a specific Greenfly Online module given the following factors:

- standards, regulations and green buildings tools such as the Green Building Council of Australia's Greenstar Interiors tool are driving rapid development in 'green' specification of office interiors products
- purchasers, occupiers, facilities managers and specifiers associated with the acquisition and use of commercial furniture are becoming rapidly sensitised to the benefits of low-impact interiors and furniture products
- designers and manufacturers of office furniture are seeking ways to undertake quick LCA and meet and exceed the requirements of standards.

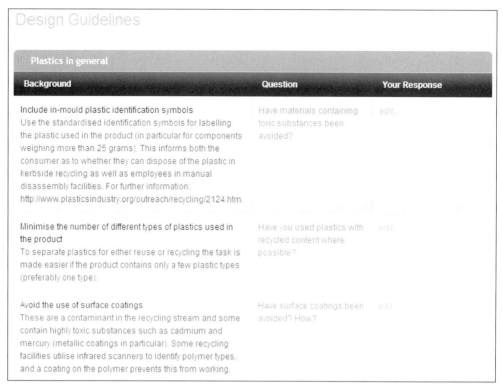

Figure 11.8   Screen shot of design guidelines from Greenfly Online generated in portable document format (PDF).

Commercial furniture is not mass-produced in the same way as many other products, but is more akin to a 'batch' system of production. This is due largely to the role of specifiers, who can specify not just products but finishes and other key aspects of products – both key to the appearance and interior design outcome and to environmental performance. Hence, a periodic LCA or a product labelling system would not meet the new requirements of specifiers to be able to establish quickly the relative environmental burdens of different types of chair coverings, desk finishes or partition materials. A Greenfly Online commercial furniture module should therefore be designed for use by specifiers as well as designers and manufacturers, and as a policy and research tool.

11.6   Conclusions

The requirement for businesses to report the environmental profile of their operations and to justify in environmental terms the materials they select and products they produce will become standard practice in the next decade, particularly with regard to greenhouse gas performance. Growing interest in LCM is a prelude to the mainstream incorporation of life cycle thinking into business processes. This will drive the need for more widespread use of LCA information. Increasing numbers of stakeholders will be involved, and more widespread LCA knowledge will develop. At the same time, the demand for speedier and easier access to information will grow rapidly. Quick LCA tools have an important role, provided they can be configured appropriately and embedded into emerging business sustainability practices. A key consideration is the balance between rigour/transparency and ease-of-use, as well as business and policy drivers and the needs of users.

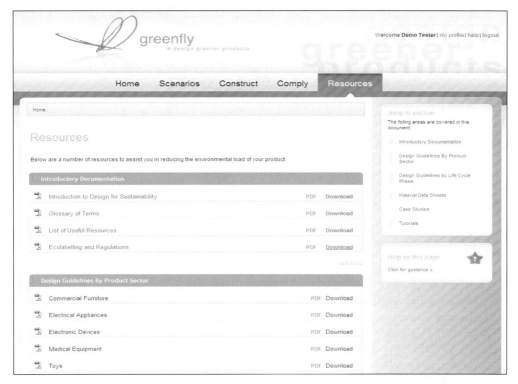

Figure 11.9   Screen shot of Greenfly Online resources page generated in portable document format (PDF).

The PIQET case study illustrates the need to consider expertise, investment of time and resources, and automation, intuitive application and assessment outcomes in quick LCA tool development. Both case studies illustrate the importance of ease-of-use, accuracy, robustness and reliability. The specific meaning and requirements inevitably vary by sector, organisation and application. Therefore, research and development must be undertaken in connection with a collaborative environment of stakeholders. Also essential are pre-existing drivers for environmental performance in the chosen sector. The champions and participants in these drivers must also be involved in the tool development. The Greenfly Online project developed from the need to solve a range of barriers to uptake and demand by both designers and their clients. Design is already information technology-intensive, so the digital platform is a good 'fit'. Also, the digital format enables easy automation of documentation and reporting, which is particularly important in communicating outcomes of eco-design processes to clients and other stakeholders. It also enables easy storage of results, ability to be searched and compared with previous and other Greenfly runs, and importantly the ability to iterate easily at a detailed level throughout the design process. The digital interface format may facilitate increased adoption because it will reinforce the problem-solving nature of the eco-design process and be easy to use. As a result, eco-design may be brought further forward in the design process.

As with many newly adopted technologies, quick LCA tools can be expected to have impacts beyond those immediately planned. By definition, unforeseen adoptions and applications are difficult to predict, but inevitably, business will look to use quick LCA information for competitive advantage in a range of ways, while regulators may look to use such information in establishing baseline performance or standards. Consumers' growing demand for clear and trustworthy environmental information can also be expected to affect the way in which quick

LCA information is used. Other applications for quick LCA tools will emerge across wider environmental management practice, research and policy development. In the final analysis, the future success of any quick LCA tool will lie in its accuracy, rigour, transparency, ease, speed and affordability, and therefore its ability to bring environmental decision-making into everyday business/design processes, as early in these processes as possible.

11.7   References

AELA (2004) 'The state of green procurement in Australia.' Good Environmental Choice Australia, Weston Creek, ACT.

Artcraft Research (2005) *Appliance Performance Labelling in Australia and New Zealand.* E3: Manuka, ACT.

Byggeth S and Hochschorner E (2005) Handling trade-offs in ecodesign tools for sustainable product development and procurement. *Journal of Cleaner Production* **14**(15–16), 1420–1430.

Centre for Design (2007) National Eco-Design Curriculum development, RMIT University, Melbourne. Retrieved 18 February 2008 from <http://www.cfd.rmit.edu.au/programs/sustainable_products_and_packaging/national_eco_design_curriculum_development>.

D'Souza C, Taghian M, Lamb P and Peretiatko R (2006) Green decisions: demographics and consumer understanding of environmental labels. *International Journal of Consumer Studies* **31**, 371–376.

Gallaroti G (1995) It pays to be green: the managerial incentive structure and environmentally sound strategies. *The Columbia Journal of World Business* **30**(4), 39–57.

Gallastegui I (2002) The use of eco-labels: a review of the literature. *European Environment* **12**, 316–331.

Gradael T (1997) Life-cycle assessment in the service industries. *Journal of Industrial Ecology* **1**(4), 57–70.

Heinrich AB and Klopffer W (2002) Editorial: LCM – Integrating a New Section in Int J LCA. *International Journal of Life Cycle Assessment* **7**(6), 315–316.

Hemmelskamp J and Brockman K (1997) Environmental labels: the German 'Blue Angel'. *Futures* **29**(1), 67–76.

Hoornweg D (1999) 'What a waste: solid waste management in Asia. The International Bank for Reconstruction and Development.' The World Bank, Washington DC, USA.

Horne R, James K, Fitzpatrick L, Jordan R and Sonneveld K (2005) Developing PIQET© – a tool for rapid packaging sustainability assessment. *Proceedings of the APRSPC Life Cycle Management Conference 2005*, Melbourne. Asia-Pacific Roundtable on Sustainable Production and Consumption, CD-ROM.

Horne RE, Wasiluk K and Lewis H (2007) 'Product environmental labels scoping study stage 1: PELS review for Sustainability Victoria.' Centre for Design, RMIT University, Melbourne.

James K (2004) 'Out of the box: Life Cycle Management in the packaging industry. How LCA will create opportunities for the packaging industry.' Paper and invited presentation to the Greening Australia and Carter Holt Harvey Breakfast Seminar, Melbourne Museum, August.

Leire C and Thidell A (2005) Product-related environmental information to guide consumer purchases – a review and analysis of research on perceptions, understanding and use among Nordic customers. *Journal of Cleaner Production* **13**, 1061–1070.

Lewis H (2006) Staying up to date on packaging environmental regulations. Sustainable Packaging Alliance, Melbourne. Retrieved 5 November 2007 from <http://www.sustainablepack.org/news/newsitem.aspx?sectionID=35andid=26>.

Lewis H and Gertsakis J (2001) *Design and Environment: A Global Guide to Designing Greener Goods*. Greenleaf Publishing, Sheffield, UK.

Lloyd S (2006) Future shock. *Business Review Weekly*, 4 May 2006, p. 38.

Lofthouse V (2006) Ecodesign tools for designers: defining the requirements. *Journal of Cleaner Production* 14, 1386–1395.

Manzini R, Noci G, Massimiliano O and Pizzurno E (2006) Assessing environmental product declaration opportunities: a reference framework. *Business Strategy and the Environment* 15, 118–134.

Monkhouse C, Bowyer C and Farmer A (2004) 'Packaging for sustainability: packaging in the context of the product, supply chain and consumer needs.' An IEEP Report for INCPEN, Institute for European Environmental Policy, London. Retrieved from <http://www.incpen.org/pages/userdata/incp/IEEPfinalreport.pdf>.

NPCC (2005) 'The National Packaging Covenant – Strategic Partnerships in Packaging. A Commitment to the Sustainable Manufacture, Use and Recovery of Packaging, 15 July 2005 to 30 June 2010.' National Packaging Covenant Council, Melbourne.

OECD (2005) 'Effects of eco-labelling schemes: compilation of recent studies.' COM/ENV/TD(2004)34/FINAL, Joint Working Party on Trade and Environment: Organisation for Economic Co-operation and Development, Paris, France.

Packaging Council of Australia (2005) 'Australian packaging: issues and trends'. Issues paper 18. Packaging Council of Australia: Retrieved from <http://www.pca.org.au>.

Pedersen ER and Neergaard P (2006) Caveat emptor – let the buyer beware! Environmental labelling and the limitations of green consumerism. *Business Strategy and the Environment* 15, 15–29.

Rebitzer G and Buxmann K (2005) The role and implementation of LCA within life cycle management in Alcan. *Journal of Cleaner Production* 13, 1327–1335.

Remmen A, Astrup Jensen A and Frydendal J (2007) 'Life cycle management – a business guide to sustainability.' United Nations Environment Programme (UNEP) and Danish Standards, Paris, France.

Rossi M, Charon S, Wing G and Ewell J (2006) Design for the next generation: incorporating cradle-to-cradle design into Herman Miller Products. *Journal of Industrial Ecology* 10(4), 193–210.

Szarka J (1991) Information failures in green consumerism. *Consumer Policy Review* 1(2), 83–86.

Taverner Research Company (2004) 'Consumer demand for environmental packaging.' Report to the NSW Jurisdictional Recycling Group, Taverner, Sydney.

Verghese K and Hes D (2007) Qualitative and quantitative tool development to support design decisions. *Journal of Cleaner Production* 15(8/9), 814–818.

Verghese K, Horne RE and Carre A (2009) PIQET: The design and development of an online 'quick' LCA tool for sustainable packaging design decision support. *International Journal of LCA* **submitted for peer review in October 2008**.

Verghese K, Horne RE, Fitzpatrick L and Jordan R (2006) PIQET – a packaging decision support tool. In: *Proceedings of the 5th Australian Life Cycle Assessment Conference*. Melbourne. Australian Life Cycle Assessment Society (ALCAS), Melbourne.

Weidema BP (n.d.) LCM – a synthesis of modern management theories. In: (2.-0 LCA consultants: Denmark). Retrieved 6 October 2008 from <http://www.lca-net.com/files/lcm9.pdf>.

Yeang K (2006) *EcoDesign: A Manual for Ecological Design*. Wiley Academy, UK.

Zaccaï E (2008) Assessing the role of consumers in sustainable product policies. *Environment, Development and Sustainability* 10(1), 51–67.

Chapter 12

Prospects for life cycle assessment development and practice in the quest for sustainable consumption

Ralph E Horne, Tim Grant and Karli L Verghese

Life cycle assessment (LCA) as a technique has been developed rapidly over the last four decades from its origins in energy accounting to international standards, global software tools and standard inventory protocols. Previous chapters have described salient aspects of this development, and charted applications in different case studies. Awareness of LCA is rising. While some users have found problems in application, such as time, resource and data difficulties, the experience and outcome is often revealing.

Inevitably, LCA development to 2020 will involve refining existing approaches and devising new applications. The need for LCA will be shaped by different events and different stakeholders, and its use and spread will be shaped by the resulting demands. In this regard, a technique is similar to a technology – it is only useful in so far as it is applied to a relevant problem, and its use is a function of the ingenuity and appropriateness with which it is applied.

Contemplating the next decade of LCA involves considering the dynamic processes in the human and natural environments that may drive the need for assessment of environmental burdens. In other words, what is the bigger picture within which LCA may take place and is LCA effective? This is discussed in Section 12.1. The next question is, given the ways in which LCA is used, how is this use likely to develop? This is addressed in Section 12.2 in the context of uptake across three sections of the economy: design and manufacturing, business management and policy making. In Section 12.3, the question posed is: What are the likely future limits of LCA? Trends are outlined and eight key themes are identified in the future of LCA practice. In Section 12.4 the problem of sustainable consumption is singled out and the extent to which LCA can contribute to its realisation in any meaningful way is questioned. Finally, in Section 12.5 there is a synthesis and conclusion with a vision for a matured, integrated LCA community of practice, with appropriate influence on decision-making towards environmentally sustainable outcomes.

12.1   Making sense of the 'problem-solution': dynamics and variables galore

Let us start by considering whether LCA 'works' – in the tradition of the 'straw man'. The application of LCA hasn't led to lowering of the eco-footprint or environmental impact of human activities; on the contrary, there is overwhelming evidence that impacts are accelerating. In Australia, existing towns and cities are spreading to accommodate a growing population and statistics suggest that Australia's population will increase to 28 million people by 2050, with continued urbanisation along coasts and around existing (already ecologically stressed) urban agglomerations (SoE 2006). Given local variations, this is not an unfamiliar picture globally. At the same time, demographic change could result in a slowing in population growth and economic growth. This may reduce pressures on the environment, although this depends on a wide range of other variables, including producer and consumer behaviours and practices.

A central difficulty in situating and gauging the effect of LCA in practice is that there are so many dynamic variables to account for in parallel. These exist in addition to the difficulties associated with data collection and impact assessment which have been considered in earlier chapters (e.g. see Chapter 5). Even where LCA studies are completed and data is sufficient for the purposes, as generally in the cases reported in waste management, the built environment and water (see Chapters 6–8), these will inevitably require further study as technologies and patterns of use change, along with energy sources.

Agricultural systems are particularly varied, as illustrated in Chapter 9. They are also in dynamic flux due to changes both in the internal structures and practices of the industry, and a wide range of external factors. The latter range from climate change to competing needs for land and raw materials, to changes in the type and quantity of demand for different agricultural products and services. In this sector especially, LCA must therefore adopt a whole-of-system view which incorporates the many implications which large scale changes in agricultural production can bring about, including scale effects, physical limits, economic ties and effect and feedback loops in the economy.

In defence of LCA, it has not really been applied as a decision support tool in any consistent, thoroughgoing way, and certainly not for very long. Where it has been applied and the results enacted, it may be expected to have led to some reductions in rates of growth of environmental impact. The problem is that, while a levelling off of rising impact may appear better than nothing, it may amount to nothing in the long run, given the rate of predicted impacts associated with climate change. In order for LCA to have a positive role in the transition towards sustainable development, we must be able to differentiate 'significant benefit' from 'negligible benefit'. Given that western consumption and resultant impacts are widely acknowledged to be unsustainable, our 'baseline' is unsustainable; how far below our current baseline lies sustainability?

So far then, we have accumulated two issues to resolve over the next decade. First, we must undertake environmental assessment and decision-making under conditions of great uncertainty and considerable change, taking account of a large number of variables. Second, we must confront the question of 'how much improvement is enough' and establish suitable orders of magnitude of expectation for environmentally preferable courses of action. Neither issue is peculiar to LCA, of course, but the conduct of LCA must be cognisant of these issues.

12.2   Future trends in LCA uptake

To look at how LCA uptake is likely to develop given current trajectories, we can take a more bottom-up, iterative approach. As indicated in Chapter 2, LCA has matured rapidly as a technique, leaving a tension between the competing needs for methodological standardisation and flexibility. Two directions are therefore possible: more detailed and complicated methods; and/or simplified and streamlined methods. We predict that both will happen, and in the following sections we contemplate how, across design and manufacturing, business management and policy making.

12.2.1   Uptake in design

While the application of LCA has spread across a wide range of manufacturing industries, products and services, there is clearly considerable scope for further uptake, especially in the frequency of use and in the earlier (design) stages of product and service development. Since most costs (both environmental and financial) are 'locked in' during design, this is an important challenge, and is also coupled with the need to make the LCA process quicker, less onerous and more accessible, while maintaining sufficient accuracy and rigour.

As discussed in Chapter 11, despite various market difficulties, some manufacturers have embraced 'Design for Environment' principles and/or LCA, across a range of sectors, from commercial furniture, mass market electrical consumer goods and even the service sector. This has led to resource efficiencies in manufacturing processes, most often following (comparative) assessment of environmental performance after design, as in 'traditional' LCA, rather than in a preventative sense, by designers. A key to unlocking the design conundrum and bringing LCA forward in the process is the development of life cycle management (LCM) tools in quick, accessible, software form readily adopted by design practitioners. With much improved databases, such tools are practicable. However, there are a range of other requirements, including the incorporation of business drivers and the integration of LCA within existing design tools across the different design disciplines (Horne *et al.* 2007). A pertinent example is the Packaging Impact Quick Evaluation Tool (PIQET) (see Chapter 11). It is also necessary to recognise the limits of the design professions to act and implement change in isolation from business, cultural, social and policy change. LCA tools need to speak equally to these disciplines if they are to be effective in supporting the aspirations of 'green' designers.

12.2.2 Uptake in business

Many businesses are already being affected by a rapidly changing set of social norms relating to environmental performance. Consumers, policy makers, regulators and other businesses are providing environmental performance challenges. The future levels of performance required and implications for a given business may not yet be known, but they will invariably require considerable attention, and can be expected to change rapidly over the next decade. There is an important role for LCA in assisting businesses in understanding their impacts and the likely challenges in reducing them. LCA will be increasingly institutionalised, being used within environmental management systems certified to the International Organization for Standardization's ISO 14001, in substantiating product environmental claims, and in driving environmental impact awareness internally and through supply chains.

Product and service providers, manufacturers and businesses have already utilised LCA, although uptake to date has been rather patchy. There is considerable scope for the further uptake of LCA in-house within businesses, to evaluate production processes and/or service provision and inform decisions about redesign, investment and development of future products and services. In particular, LCAs of products/services will become more systematic as attempts gather pace to drive down supply chain environmental burdens and reduce carbon emissions and related costs.

Such embedding processes provide the opportunity to develop or expand in-house LCA awareness and capacity. A useful 'goal' for business managers might be to make environmental evaluation and decision support techniques such as LCA as ubiquitous and well understood as financial management techniques, and to make LCA data at least as widely understood and available as economic statistics, and financial reports and accounts. This will make LCA information cheaper to obtain and use, and will also facilitate transformation of decision-making through the inevitable seismic shift in social capacity resulting from widespread knowledge and practice involving the exchange of information about life cycle environmental burdens.

12.2.3 Uptake by policy makers

The point about making environmental data and LCA practice as ubiquitous, accessible and widely understood as economic and financial practices and data is also highly pertinent to policy makers. In addition, there is the prospect of incorporating LCA data or techniques into the regulatory processes themselves. Previously this has been rather fraught with difficulty due to the methodological issues of LCA and data needs as discussed throughout this book. The Dutch experience with building materials and LCA (see Chapter 7) indicates that policy development should proceed carefully in considering such developments.

However, there are solid grounds for wider use of LCA in informing policy development. Indeed, building materials provides a further example here, of a sector where future policy in Australia is now being informed by LCA. This raises the question: what should any new policy or regulation influenced by LCA studies be expected to achieve? Most critically, the attention must be focused on preventing the penetration of poorly performing goods and services into markets at artificially low prices. These low prices are achievable only because the producers do not pay the full costs of production; they simply pay labour and materials, and the costs associated with, for example, pollution and human health, remain as 'externalities' to be paid by society as a whole. Even where the producer pays generic taxes, these are invariably uncon-nected to the externalities, and so it is often in the producer's interest to maximise external (environmental) costs in order to reduce internal costs and maximise profit. The problem of environmental externalities has been long debated, and critiques of neo-classical approaches indicate clearly that our current economic systems are simply inadequate as a basis for 'full costing' of environmental impacts (e.g. Horne 2001).

Although there has been an active debate in Australia, in which the Productivity Commis-sion has repeatedly argued that external environmental costs are not sufficient to warrant intervention to fix market distortions, consumers are only happy, in the main, to pay the 'full' cost by buying environmentally benign goods, when they know their peers will not be able to 'free-ride' by buying cheaper, poorly performing goods, while everyone suffers the environ-mental consequences. For example, Zaccaï (2008) suggests that consumers are not convinced of the importance their contribution could have and continue, mainly, to expect a legislative setting from public authorities (see Section 12.4). This issue has now become a moral as well as an economic one, and policy makers will be able to utilise LCA in developing fuller under-standing of the environmental metrics when contemplating appropriate policies and regula-tions to address what constitutes a wide range of current market distortions.

12.3   Beyond comparison? A quick guide to LCA development to 2020

Following on from the trends and likely directions of LCA, we come to the question of limits: What are the outer boundaries of LCA, and what other questions around environmental assessment will remain unanswered by application of the technique? What will shape the direction of LCA over the next decade and what should users and practitioners be aware of? Eight key points are made here as a quick guide.

There is a need for wider LCA awareness, so that neither LCA practitioners nor lay recipi-ents of LCA information are tempted to view LCA results as definitive or absolute. This creates a heady and potentially dangerous mix, as LCA results may often appear profound (especially to the lay person who has little other information to go on), and the temptation is often to apply a single LCA result well beyond its intended or appropriate limits. Add to this the other temptation of commercial organisations to seek the 'right' answers for their products' environ-mental performance by tweaking LCA methods and results, and it is unsurprising that LCA is sometimes viewed as controversial. Nevertheless, given the choice between acceleration of the breadth of application of LCA and a 'new repression' through risk reduction and control of applications, we favour more adventure in LCA and environmental assessment rather than less; but with the foresight of reflexive practice and measured experience. In accepting this challenge, we can now consider what experience can tell us about where LCA may venture – and what should be regarded as off-limits.

So, the first point must be that the interpretation of LCA results should not be undertaken lightly, and the special responsibility that LCA practitioners have as stewards of rare insights into the environmental flows of human processes (and by inference, directions for policy) should not be underestimated. An LCA may reveal the world behind a product, but is this sufficient to direct new product processes, or to reveal the relative costs and benefits of a world without that product?

Given the vast variability and dynamics of ecological, cultural, economic, legal and political structures and systems as indicated above, we must recognise that there is an interdependent relationship between the development and provision of environmental impact information and the resulting practice and policy responses. Developments in policy contexts across different sectors and the growth of LCA-related stakeholder groups are examples of a growing soft infrastructure and awareness concerning life cycle impact concepts. As these develop, tensions arise within LCA practice as to appropriate directions and developments, as the political context manifests itself. Institutionalisation of LCA could be regarded as a means by which control is exerted over the application of the technique, therefore influencing the results obtained, whether through direct influence over data or, more likely, over careful manipulation of the questions and system boundaries.

Second, like a toddler gaining confidence in walking, LCA will become more purposeful and surefooted as it expands in application. For example, input-output based LCA has already provided useful indications of patterns and trends of agricultural impacts, and more specific studies have been undertaken (e.g. see Chapter 9, Foran *et al.* 2005; Wood *et al.* 2006). In the future, process LCA application can be expanded significantly to: address outstanding questions around future agricultural efficiency and security; investigate the impacts associated with different scales of agriculture (e.g. small, medium and large farms, urban crops); examine optimal packaging and storage systems to control food wastes for least overall impact; and identify combined environmental and health impacts and benefits of different organic and low input, and other alternative agricultural practices and systems.

Third, in concert with increased application is the need for improvements in the technique – particularly in methods and data. Indeed, there remains a multitude of needs for environmental data arising through energy, water, greenhouse and pollution reporting schemes, including voluntary reporting using LCA data by businesses. This has been indicated at many points throughout this book and so need not be expanded further here, except to note that there are already data inventory development projects underway, including the Australian inventory project led by ALCAS (see Chapter 2), and there is emerging potential to mine the Internet for LCA data from manufacturers and suppliers.

The use of special Internet mark-up language that identifies LCA data is useful because web-crawling tools can find and organise the data in the same way that search engines organise other Internet content. One open source database initiative is the Earthster initiative, which is building life cycle tools for assessing the impact of purchases. Initially, the project is based on public databases, but there is a longer term plan to develop tools to source other Internet-based data sources as potential suppliers, so that anyone can compare product LCA performance with the average documented in LCA databases (Norris 2007). Apart from resources and time, issues of third-party validation and commercial confidentiality are key issues which such initiatives are dealing with.

Fourth, LCA has already proved its worth in indicating that initiatives based on single issues or partial analyses, whether food miles, packaging or energy use, may indicate legitimate issues, but the process has shifted the risk burden and increased impacts overall. LCA is the appropriate technique to minimise these risks, through its systematic, multi-impact approach, and it can be expected to be invoked repeatedly to shed light on the 'total' impact of products and services over the coming decade. For example, the rapid rise in interest in greenhouse gas emissions in Australia since 2006 is a significant catalyst for increased LCA uptake, including to inform 'carbon trading' regimes and to indicate the wider environmental impacts of 'carbon management' based initiatives.

The fifth point is that LCA can be extended beyond its 'natural' limits, and therefore strengthened contextually, by integration and comparison with other assessment approaches. For example, since LCA is systems and functional unit based, it does not lend itself easily to issues of carrying capacity or scale. Hence, impacts are typically considered on a unit, rather than a bulk or absolute, basis. Other tools which do focus on macro-economic or environmental stocks and flows can therefore be used together with LCA to provide uniquely illuminating results. An example is the use of the Commonwealth Scientific and Industrial Research Organisation (CSIRO) Australian Stocks and Flows Framework with LCA in a recent study on building materials (DEWR 2007, see Chapter 7).

Similarly, LCA is typically used in situations which are place-independent. Where spatial issues are significant, it can be applied with tools such as Environmental Impact Assessment, to combine the rigour of systematic, function-based assessment with site-specific assessment techniques.

Integrated assessment also holds out the possibility of more clearly juxtaposing alternative sustainability perspectives and evaluations. For example, it has been suggested that there are four distinctive approaches to environmental sustainability assessment:

1. Industrial ecology approaches with a mass orientation, as in MFA (Materials Flow Analysis) and SFA (System Flow Analysis) (Brunner and Rechberger 2003).
2. Technology oriented life cycle approaches, exemplified in ISO-LCA (ISO 14040 series) and EIOA (Environmental Input-Output Analysis) (Tukker et al. 2006; Huppes et al. 2006; Leontief 1970).
3. Main stream economists taking into account market relations, exemplified in CBA (Cost Benefit Analysis) and CGE (Computable General Equilibrium) modelling (Barbier et al. 1990; Eshet et al. 2006; E3ME 2008; GEM-E3 2008).
4. Ecological economists, refraining from general approaches, focusing on multi-criteria analysis at a case level, and therefore lacking an acronym (Martinez-Alier et al. 1998) (Huppes and Ishikawa 2007, p. 62)

While the authors do not suggest that integration would reconcile different results or values that each approach embodies, they point to the value of acknowledging a taxonomy of sorts, and support a Society of Environmental Toxicology and Chemistry (SETAC) Working Group initiative to clarify the relations between these overlapping approaches.

The sixth point extends the integration theme: LCA penetration can be improved through the development of more accessible tools and languages relevant to relevant sectors, professions and practitioners. For example, there is scope for the development of functional and integrated computer-aided design (CAD)-LCA tools for different sectoral applications, to enable CAD users to more easily undertake and incorporate LCA information into their design activities. Such initiatives are already underway (e.g. in Australia, the CRC for Construction and Innovation Initiative LCA-Design). Also, as indicated in Chapter 11, LCA methods will increasingly become part of assessment and design tools for specific applications. Examples where this is already happening are in packaging design, solid waste management and building assessment. Typically, the tools model the technical system under consideration such as the product, building or technology configuration so that the material requirements and operational energy can be predicted and interpreted by the LCA. There is more potential to develop LCA in this way in the following sectors:

- water services planning
- product design specialties
- purchasing tools.

The seventh point is related to the issue of easing accessibility to LCA information, but extends to include the development of capacity to use this information. There is a need to provide significant additional capacity for academic, professional and in-service training, including through embedded modules within planning, design, and built environment disciplines, and well beyond, across social sciences and engineering. Engineering is the only discipline to currently incorporate notable curricula elements which could be included under the term LCA in any significant way, and there is an urgent need for other disciplinary contexts for LCA to be explored.

The need for increased capacity also extends beyond 'training', since there is a need not just for a new profession of LCA 'experts' but also (and perhaps more fundamentally) a need for LCA capacity to be embedded across disciplines, organisations and policymaking. Despite growing awareness and access to available information about the impacts of climate change, our current processes of learning have not always translated into action. Programs of wider education and awareness raising to develop active engagement, knowledge and ownership of LCA as a valuable tool in assisting society along a more sustainable path will assist in this if they can be participatory and based in practice.

Social learning (e.g. Wals 2007) highlights the transformative nature of participation and 'ownership' in learning and outcomes effectiveness. While observation and 'traditional' training and information helps establish and maintain norms for accepted and expected decisions, actions and behaviour (Myers 2007), this gathers more potency when more participative learning environments are created, such as the 'use of group gatherings and processes such as workshops and meetings where ideas, knowledge, experiences, skills and practices may be exchanged' (Voronoff 2005).

There are clear implications here for the ways we may envisage the methods and propagation of LCA capacity development. Glasser (2007) differentiates between passive and active social learning. Passive learning does not rely on interaction. It includes newspapers, books, lectures, the Internet and direct observation and while it can provide useful new insights it 'generally has limited applicability for directly spawning substantively new social innovations' (Glasser 2007, p. 50). Active social learning, however, involves 'conscious interaction and communication between at least two living beings' and is 'inherently dialogical' (Glasser 2007, p. 51). He contends that most learning in our society is passive, and can and does continue to perpetuate 'maladaptive' processes; for example, the maintenance of practices and institutions that are harmful to the environment. He further differentiates between hierarchical, non-hierarchical and co-learning, the latter being based on collaboration, trust, full participation, and shared exploration, and holding out the most potential for transformative, lasting action.

Finally, there is a need to recognise the limits of LCA and to probe these limits through research, while at the same time ensuring that practice remains within them. LCA is typified by the comparison of two or more systems in order to 'pick the best', and the initial problem is: what if neither are any use? While LCA practitioners are getting more adept at identifying from a range of alternatives which has a potentially lower impact (notwithstanding the data and methodological challenges discussed through previous chapters), the application of LCA to wider questions of sustainability remains constrained in many regards. Some such constraints are indicated in Chapter 4, and the framing problem – the complexity and variability of environmental assessment itself – is introduced in Section 12.1.

Perhaps the essential concern with options-based assessment is that it can lead the unwary to frivolous, ill-advised, or simply unsustainable ends. An LCA comparing three different routings for a new freeway may direct the debate away from whether a freeway is a good option at all, or whether it is possible to construct and operate one within carrying capacity limits. The classic product comparison rarely compares against non-consumption options as, arguably, this would not be in producers' interests (or those of a free market economy driven by inexorable growth in consumer spending). Thus, LCA is generally concerned with the supply side of environmental impacts rather than the demand side. Can it help us achieve sustainable consumption, rather than simply identifying the 'less-than-worst' options?

12.4   LCA and sustainable consumption

Unsustainable consumption dates back to pre-Christian civilisations, and even since the industrial revolution, consumption concerns date back to the mid-19th century. As Tim Jackson (2006, p. 2) points out, regarding sustainable consumption: 'The terminology is recent. But the concern with resource consumption is scarcely new'.

In the modern and divergent discourse of sustainable consumption between 'consume efficiently' at the accommodation end and 'change lifestyles' at the paradigm shift end, where might LCA fit? Lifestyle and behaviour change are now mantras of policy makers in the face of the realisation that tinkering with technologies (such as has been achieved) is not enough to prevent dangerous climate change. Everything from food choices to energy saving behaviours to transport modes and leisure pursuits are candidates for a new phase of social engineering in the face of environmental necessity. Can LCA play a role here?

Presently, LCA is often associated with efficiency. However, there are many levels and uses of the term. An early problem arises when efficient choices or rational actions are assumed. Application of LCA can lead to cleaner production where greater efficiency is indicated, and where this is related to costs in a competitive marketplace, and where actors are empowered, resourced and willing to follow greater efficiency goals. However, we are not always rational – and we exhibit different rationalities at different times. Moreover, as consumers, we often do not rationalise between products or services, we just 'do' – Shove's work illustrates that consumption practices are likely to be more dominated by a series of social practice determinants which sit 'below the radar' rather than by conscious rational decision-making (Shove 2003a, b; 2006).

As indicated in the literature (e.g. OECD 2005) (see Chapter 11), many people view themselves as green consumers, but relatively few act consistently green. Clear labelling backed by independence of claims and public information campaigns help, but it remains that consumption and purchasing are complex - and green purchasing is no less so. Many consumers do not trust the information they do get and, while people often state a preference to consume more sustainably, convenience, comfort, cost and a vast range of cultural and social norms and practices often appear to conspire against us when it comes to following through. Moreover, when consumers do take the 'green' option, they often appear to compensate for this by then consuming elsewhere – either consciously, rather like a chocolate reward for careful dieting that backfires, or simply out of convenience, ignorance or other reason. Also, where the choice is to buy green or non-green, as indicated in Section 12.2, consumers are less likely to 'pay for green' if others are free-riding in an imperfect marketplace.

So, among consumers, key issues in the uptake of green goods include price, awareness, trust, the complexity and availability of information, and the interplay of this with other behavioural and social factors. Given such a range of influences, under what conditions do we switch from a high impact to a low impact product or service, and when can LCA safely assume there will be no rebound effect? Moreover, what may prompt consumers to choose not to purchase or use a product? What would this mean for functional units of comparison? While literature exists on well-being and stated preferences, more intellectual bridges need to be built between the LCA or impact assessment literature and the behavioural and sociological literature.

With respect to functional units in LCA, products or services are typically compared on the basis of some proportional or comparable service outcome. Thus, two tables may be compared on the basis of 'ability to support, per metre squared area, per year of design life'. However, comparing two tables by size says little about their function as representations to ourselves and others. In properly seeking to compare services rather than things, the true value in terms of well-being or quality of life outcomes of that service must be known, on a suitably comparable scale. What LCA cannot afford to do is follow the flawed utilitarian logic that utility can be measured by the market or by acquisition of more goods, since there is little correlation between increased income and reported well-being (Ingelhart and Klingemann 2000). Lapham (1988) concludes from extensive research in the United States of America (USA) that no-one ever has enough: typically they require a doubling of salary to make them comfortable (irrespective of their current salary). Only by valuing a product or service in terms of 'needs met' can comparisons be made on a like-for-like basis – yet there is an entire literature illustrating that the line between need and desire is blurred (e.g. Princen *et al.* 2002). Armed with an increasing awareness of the dynamic and moveable nature of functional utility, LCA practitioners and researchers may be increasingly able and willing to experiment with different 'functional units' scenarios in LCA research.

Undoubtedly, sustainable consumption provides fundamental challenges to society; which extend well beyond the scope of any single technique. A multitude of interventions is required, and LCA can contribute – for example, there are opportunities for developing more direct consumer-LCA information interaction. This may occur through established eco-label systems of various sorts, or through new initiatives that enable consumers to access clear, transparent and reliable LCA information without creating information overloading. For example, using mobile digital interfaces integrated with barcodes, LCA information could be as instantly available as price at the supermarket shelf. It is an indication of the pace of change both in technology and social norms that while such developments would have been widely considered to be unlikely in 2005, they now appear quite likely well inside the 2020 timeframe.

12.5   Conclusion: towards reflective, integrated practice

LCA can make a contribution to one of the greatest challenges of modern civilisation: the transition to sustainable consumption. It is no secret that significant shifts are required – in technologies, in decision-making and in thinking. Since the industrial revolution we have tied our endeavours to extracting more materials and energy; now we have to learn how to leave fossil fuels in the ground (at least until we learn how to capture carbon economically). At a time when society struggles with the need for a paradigm shift, LCA is rapidly maturing as a powerful technique for revealing the environmental loads that lie behind our products and services.

LCA has faced a number of challenges as a relatively new technique with significant data needs and other facets that require it to obtain 'buy-in' from a wide range of stakeholders and professionals, including designers, technicians, managers and policy makers. The unique approach of LCA – and its tendency to produce counter-intuitive and even controversial results and new perspectives – is both a strength and a challenge. The positive outcome is increased exposure and recognition of the value of LCA, while the problems arise from opposition to the results – often from stakeholders who have a vested interest in maintaining the status quo.

Within a sea of variables and rapid shifts – in everything from climate to culture – LCA has established a systematic, rigorous and revealing approach that often sheds new light on the environmental implications of our endeavours. This is achieved through a range of assumptions and truncations which, while enabling the technique, may also devalue it if these are not explicit. In the preceding sections, the authors have indicated likely developments and uptake in LCA, and developed eight themes which, taken together, are designed to minimise the risk of mistakes, and maximise the benefits of the application of the technique. These can be summarised as:

1. proceed boldly but with care
2. practice will improve effectiveness
3. continual improvement is necessary
4. bring breadth to 'single' issues
5. use in conjunction with other techniques
6. improve ease of access
7. build capacity and embed practice
8. recognise limitations and engage with other disciplines.

In general, these themes need no further explanation since they have been developed already; however, the last two themes are possibly the most pressing and complex, and warrant further emphasis.

The use of social learning approaches could be expected to assist in extending the capacity for and use of 'life cycle thinking' both institutionally and among the wider public and social realms. This is bound up with the propagation of LCA terminology, information and findings in popular culture and a range of public settings. The emerging social norms of 'good environmental citizenship' and their interplay with the transition from novel applications of LCA to 'mainstream' common usage may be expected to grow over the next decade, so that, by 2020, it will be as common to overhear conversations in pubs over environmental impacts associated with a trip to the 'footy' as about the associated cash costs and logistics.

An opportunity also exists to expand the settings in which LCA is applied to help bridge the gap between the technical disciplines and the policy and social realms. This is particularly stark in the realm of sustainable consumption. While LCA is inherently limited in the wider realm of sustainable consumption concerns, it also benefits from them, since these have led governments to examine material and energy flows; in other words, to use LCA as a technique to shed light on the environmental impacts of household consumption (e.g. see Hertwich 2005). Increasingly, development of traditional environmental indictors is accompanied by socioeconomic and social indicators in LCA (e.g. there is a United Nations Environment Programme (UNEP)/SETAC working group on social indicators). As the boundaries of LCA continue to expand, the potency of this technique will be maximised through the growing community of LCA practitioners pursuing open, reflective and increasingly integrated scholarship and practice.

12.6   References

Barbier E, Markandya A and Pearce DW (1990) Environmental sustainability and CBA. *Environment and Planning A* **22**, 1259–1266.

Brunner PH and Rechberger H (2003) *Practical Handbook of Material Flow Analysis.* Taylor and Francis/CRC, London.

E3ME (2008) Cambridge Econometrics. Retrieved 3 May 2008. <http://www.camecon.com/e3me/e3me_model.htm>.

Eshet T, Ayalon O and Shechter M (2006) An inclusive comparative review of valuation methods for assessing environmental goods and externalities. *International Journal of Business Environment* **1** (2), 190–210.

Foran B, Lenzen M and Dey C (2005) *Balancing Act – A Triple Bottom Line Analysis of the Australian Economy.* CSIRO, Sydney.

GEM-E3 (2008) Retrieved 3 May 2008, <http://www.gem-e3.net>.

Glasser H (2007) Minding the gap: the role of social learning in linking our stated desire for a more sustainable world to our everyday actions and policies. In: *Social Learning Towards a Sustainable World: Principles, Perspectives and Praxis.* (Ed. EJ Arjen Wals) Chapter 1. Wageningen Academic Publishers, Wageningen, The Netherlands.

Hertwich EG (2005) Lifecycle approaches to sustainable consumption: a critical review. *Environmental Science and Technology* **39**, 4673–4684.

Horne RE (2001) Regulating the environmental impacts of the electricity supply industry. PhD Thesis, Sheffield Hallam University, UK.

Horne RE, Wasiluk K and Gertsakis J (2007) Rapid life cycle assessment design tools and their role in DfE transitions in Australia. In: *Proceedings, 5th International Conference on 'Design and Manufacture for Sustainable Development',* Loughborough, UK, 10–11 July 2007. CD-ROM.

Huppes G and Ishikawa M (2005) A framework for quantified eco-efficiency analysis. *Journal of Industrial Ecology* **9** (4), 25–41.

Huppes G and Ishikawa M (2007) Sustainability evaluation: diverging routes recombined? Tasks for a new working group on modelling and evaluation for sustainability. *International Journal of Life Cycle Assessment* **12** (1), 62.

Huppes G, Suh S, Heijungs R, van Oers L, Nielsen P and Guinée JB (2006) Environmental impacts of consumption in the European Union: High-resolution input-output tables with detailed environmental extensions. *Journal of Industrial Ecology* **10**(3), 129–146.

Inglehart R and Klingemann H-D (2000) *Genes, Culture and Happiness.* MIT Press.

Jackson (2006) *The Earthscan Reader on Sustainable Consumption.* Earthscan, London.

Lapham LH (1988) *Money and Class in America: Notes and Observations on Our Civil Religion.* Weidenfeld and Nicolson, New York.

Leontief W (1970) Environmental repercussions and the economic structure: an input-output approach. *Review of Economics and Statistics* **52** (3), 262–271.

Martinez-Alier J, Munda G and O'Neill J (1998) Weak comparability of values as a foundation for ecological economics. *Ecological Economics* **26**, 277–286.

Myers DG (2007) *Psychology.* 8th edn. Worth Publishers, New York.

Norris G (2007) 'Green Guide to Green Build – LCA Impact: Life Cycle Assessment: Out in the Open. A look at how LCA information can become more accessible, transparent and dynamic.' Interiors and Sources: Green Guide. October/November. Statmats Business Media, Cedar Rapids, Iowa, USA.

OECD (2005) 'Effects of eco-labelling schemes: compilation of recent studies.' COM/ENV/TD(2004)34/FINAL, Joint Working Party on Trade and Environment: Organisation for Economic Co-operation and Development, Paris.

Princen T, Maniates MF and Conca K (2002) *Confronting Consumption.* MIT Press.

Shove E (2003a) *Comfort, Cleanliness and Convenience: The Social Organization of Normality.* Berg, Oxford and New York.

Shove E (2003b) Users, technologies, and expectations of comfort, cleanliness and convenience. *Innovation* **16** (2), 193–206.

Shove E (2006) Efficiency and consumption: technology and practice. In: *The Earthscan Reader in Sustainable Consumption*. (Ed. T Jackson) p. 293. Earthscan, London, UK.

SoE (2006) 'Australia State of the Environment Report 2006'. Commonwealth of Australia, Canberra.

Tukker A, Eder P and Suh S (2006) Environmental impacts of products: policy relevant information and data challenges. *Journal of Industrial Ecology* **10**(3), 183–198.

Voronoff D (2005) *Community Sustainability: A Review of What Works and How It Is Practiced in Victoria*. Environment Victoria, Melbourne.

Wals AEJ (Ed.) (2007) *Social Learning Towards a Sustainable World: Principles, Perspectives and Praxis*. Wageningen Academic Publishers, Wageningen, The Netherlands.

Wood R, Lenzen M, Dey C and Lundie S (2006) A comparative study of some environmental impacts of conventional and organic farming in Australia. *Agricultural Systems* **89**, 324–348.

Zaccaï E (2008) Assessing the role of consumers in sustainable product policies. *Environment, Development and Sustainability* **10**(1), 51–67.

Index